MODERN WORLD NATIONS

AFGHANISTAN
ARGENTINA
AUSTRALIA
AUSTRIA
BAHRAIN
BERMUDA
BOLIVIA
BOSNIA AND HERZEGOVINA
BRAZIL
CANADA
CHILE
CHINA
COSTA RICA
CROATIA
CUBA
EGYPT
ENGLAND
ETHIOPIA
FRANCE
REPUBLIC OF GEORGIA
GERMANY
GHANA
GUATEMALA
ICELAND
INDIA
INDONESIA
IRAN
IRAQ
IRELAND
ISRAEL
ITALY
JAMAICA
JAPAN
KAZAKHSTAN
KENYA
KUWAIT
MEXICO
THE NETHERLANDS
NEW ZEALAND
NIGERIA
NORTH KOREA
NORWAY
PAKISTAN
PERU
THE PHILIPPINES
RUSSIA
SAUDI ARABIA
SCOTLAND
SENEGAL
SOUTH AFRICA
SOUTH KOREA
TAIWAN
TURKEY
UKRAINE
UZBEKISTAN

MODERN WORLD NATIONS

Indonesia

Douglas A. Phillips

Series Consulting Editor
Charles F. Gritzner
South Dakota State University

CHELSEA HOUSE
PUBLISHERS
A Haights Cross Communications Company
Philadelphia

Dedication: This book is dedicated to my parents, Carolyn and Alan M. Phillips. Their love and belief in me has always provided me with the element of hope. Their guidance and mentoring fostered my quest for travel, exploration, and learning. From outhouses to the Outback and from homemade travel trailers to the world, my parents opened my eyes to the worlds of education and travel. My love, admiration, and gratitude can never be expressed by words.

Frontispiece: Flag of Indonesia

Cover: Rice barns for storing grain in Bukit Lawang, Sumatra.

CHELSEA HOUSE PUBLISHERS

VP, New Product Development Sally Cheney
Director of Production Kim Shinners
Creative Manager Takeshi Takahashi
Manufacturing Manager Diann Grasse

Staff for INDONESIA

Executive Editor Lee Marcott
Production Editor Noelle Nardone
Photo Editor Sarah Bloom
Series Designer Takeshi Takahashi
Cover Designer Keith Trego
Layout 21st Century Publishing and Communications, Inc.

A Haights Cross Communications Company

http://www.chelseahouse.com

First Printing

1 3 5 7 9 8 6 4 2

Library of Congress Cataloging-in-Publication Data

Phillips, Douglas A.
Indonesia / Douglas A. Phillips.
p. cm.—(Modern world nations)
Includes bibliographical references and index.
ISBN 0-7910-8022-6
1. Indonesia—Juvenile literature. I. Title. II. Series.
DS615.P55 2004
959.8—dc22

2004012759

Table of Contents

1	Introducing Indonesia	8
2	Natural Environment	14
3	Indonesia Through Time	24
4	Independence and Nationhood	36
5	People and Culture	46
6	Government of Indonesia	62
7	Indonesia's Economy	74
8	The Islands	84
9	Indonesia's Future	92
	Facts at a Glance	98
	History at a Glance	100
	Further Reading	103
	Index	105

MODERN WORLD NATIONS

Indonesia

CHAPTER

1

Introducing Indonesia

It is hard to imagine Indonesia as a country. Thousands of islands scattered across the expanse between the Indian and Pacific Oceans is hardly a likely candidate for a unified country. Yet this is the land (and water!) of Indonesia. It is a country pieced together from more than 17,000 islands, more than 6,000 of which are inhabited. The arrangement of Indonesia is much like an island splatter pattern running from east to west and located off the coast of Southeast Asia. The tyranny of distance created by the water separating these islands adds many complexities for the people and government of Indonesia. These challenges will be examined in this book.

Indonesia often brings to mind visions of a tropical island paradise, with palm trees swaying in the warm breeze and sandy beaches that seem to stretch endlessly toward the sea and a picturesque sunset. When the island of Bali is mentioned, it is often in the context of a dream destination for travelers. In recent years it has been a frequent

Indonesia is a country composed of thousands of islands scattered off the coast of Southeast Asia. Although it has the fourth-highest population in the world, it is actually only about one-fifth the size of the United States.

destination for travelers from Australia and the West. Sadly, this dream was partially shattered on October 12, 2002, when a terrorist's bomb created a nightmare for travelers and the Indonesian government. This happened when the Sari Club was bombed on the island of Bali; the perpetrators were linked to al Qaeda. More than 200 lives were lost in the blast, a murderous act that now threatens to damage the vitality of Indonesia's tourism industry. Has the idealistic vision of Indonesia and Bali been ruined forever by terrorists?

How did al Qaeda's tentacles reach into Indonesia? What exactly are Islamic extremists doing in paradise? Indonesia is a country vastly more complex and troubled than the image presented by the simple vision of an island paradise on Bali. Few Westerners recognize that Indonesia, with more than 230 million people, ranks fourth among the world's countries in population. In addition, a staggering 88 percent of its people, more than 200 million, are Muslim. This is about the same number of Muslims as the nations of Egypt, Iraq, Iran, Saudi Arabia, Syria, Lebanon, and Jordan combined. In other words, Indonesia has nearly the same number of Muslims as the entire Middle East.

Although Islam is Indonesia's dominant faith, the country works to protect religious freedom for its citizens. Residents are also Roman Catholic, Hindu, and Buddhist; they also practice a number of Protestant denominations, plus a smattering of other religions.

Indonesia's peoples are diverse; regional interests and cultural elements vary widely. According to the U.S. Library of Congress, more than 650 different languages and dialects are spoken in the country. The partition of Indonesia by water has often made interisland communication and transportation difficult. Because they lived in relative isolation from others, residents of different islands developed richly unique cultures. Thus, local languages such as Javanese, Balinese, Sasak, and Sundanese are still used in some areas, although Bahasa Indonesia serves as the national and unifying language. Bahasa Indonesia is a variation of the Malay language. English and Dutch are also frequently spoken, because Indonesia was once a colony of the Netherlands, and English has become the international language of business.

Indonesia's population ranks behind only that of China, India, and the United States, but a major difference between

Indonesia and these other four populous nations is that its land area is significantly smaller. With a land area of 705,192 square miles (1,826,440 square kilometers), Indonesia is about three times the size of Texas, or a little more than one-fifth the size of the United States or China. This means that the population density of this tropical country is severe in many locations suitable for settlement. Jakarta, the capital city, is home to some 11 million people who dwell in its metropolitan area. As a country, the population density exceeds 300 people per square mile (190 per square kilometer) compared to the just under 80 per square mile (50 per square kilometer) density in the United States. Indonesia's population density figure is very deceptive, however, because much of the country's land remains isolated or uninhabitable. Most of the population lives near the coast, where soils are more fertile and land is less rugged. On many of the larger islands, a spine of mountains runs down the length of the land mass. This ruggedness makes much of the land difficult to live on or to use for agriculture. Other islands suffer from their remote location and isolation, a condition that makes living on them a challenge.

Indonesia is located between the Indian Ocean (to the west) and Pacific Ocean (to the east). It comprises the world's largest island chain, an archipelago laced in and around the equator, and stretches for nearly 3,000 miles (4,800 kilometers), from Sumatra in the west to Irian Jaya in the east. Indonesia is located entirely within the tropics. Yet some of the country's higher peaks are blanketed with snow and ice throughout the year. Its closest neighbors are the new country of East Timor (formerly a part of Indonesia) and the countries of Malaysia and Papua New Guinea.

Indonesia's location on the Pacific Ocean's "Ring of Fire" makes it susceptible to frequent earthquakes and violent volcanic activity. It has experienced the most violent eruptions in recent history. This and other aspects of the country's physical setting will be discussed in chapter 2.

The unrest in Indonesia is not just geological—much political, social, and economic turmoil exists within the country, as well.

These differences create major challenges for the government of the Republic of Indonesia (the country's official name). The Indonesian government has three branches, but it has only a one house, or legislative chamber, called the People's Representative Council. The president mainly comprises the executive branch, and the Supreme Court heads the judiciary. With the large Muslim population, the law is sometimes complex because Sharia law, moral law from the most holy book of Islam, the *Koran*, often plays an important role in local courts.

The economy of Indonesia has changed remarkably since 1965, when little industry existed in the country. Today the country boasts a mixed economy, with a variety of activities, including agriculture, manufacturing, and the extraction of natural resources such as oil and natural gas. Indonesia's economy suffers, however, because of political instability in various regions of the country. In a number of areas, for example, secessionist activities make outside economic investment precarious. In addition, the cancer of corruption has had an insidious effect on the country's government and economy.

Indonesia helped its economic development by serving as one of five founding members of the Association of Southeast Asian Nations (ASEAN). This regional organization was created by the ASEAN Declaration, signed in Bangkok, Thailand, on August 8, 1967, by Indonesia, Malaysia, the Philippines, Singapore, and Thailand. Today 10 countries belong to ASEAN. (Cambodia, Brunei Darussalam, Laos, Burma [Myanmar], and Vietnam have also become full members.) Countries in ASEAN strive for economic integration, but the organization also promotes cultural, educational, technical, and social relations as well as political stability in the region.

At the beginning of the twenty-first century, Indonesia is in a predicament. The natural environment creates unexpected explosions that threaten people's lives and disrupt daily life—the precarious nature of living on the Pacific Rim and the Ring of Fire. Violent elements from al Qaeda and local separatists in

Indonesia's position in the Pacific Ocean's ring of fire means there is a constant threat of earthquakes and volcanic eruptions. This volcano in the Bukit Barysan Mountains of Sumatra is one of many active volcanoes throughout the islands.

areas such as Aceh have intruded on the human environment in recent years, making it, too, more explosive. Great opportunities exist for Indonesia, however, because of the natural and human resources available for this country's development. This book will take you on an unparalleled journey from cities (such as Jakarta) teeming with people, to remote islands with isolated populations. During the journey, you will visit environments as varied as the violent volcano of Tambora and the calming beaches of Bali. We will examine the paradoxical nature of Indonesia and at the same time will take you on a trip that will be more exciting that just island hopping across the vast archipelago. It will be the adventure of a lifetime.

CHAPTER

2

Natural Environment

Indonesia is an archipelago (chain of islands)—the world's largest—with some 17,000 islands scattered in an east–west distance of more than 3,000 miles (4,800 kilometers). It is also a tropical land, with no part of the country lying more than 11 degrees of latitude from the equator. As you might imagine, it is very difficult to determine precisely the country's area. Figures vary by more than 50,000 square miles (130,000 square kilometers)! Regardless, the land area is roughly comparable to that of the United States east of the Mississippi River—even though it lies scattered over a vastly greater area.

Indonesia's tropical location might suggest a rather uniform landscape, but the country offers many surprises, as you will learn in this chapter. By any measure, it is a tropical "paradise" yet has snowcapped mountain peaks, is home to perhaps the world's greatest diversity in plant and animal life, and is a land subject to some of the Earth's most devastating natural elements and events.

LAND FEATURES

Indonesia, it can be said, has been both blessed and cursed by the Earth's geologic forces. It is blessed by the beauty of countless tropical islands scattered like gems across a mirrorlike sea. Its terrain is varied, with many towering mountains breaking the monotony of broad and fertile lowland plains. The shimmering snow-capped peak of Puncak Jaya towers 16,503 feet (5,030 meters) above the steaming tropical coastal plain of New Guinea. Most Indonesians live on fertile plains, where farming, building, and access by road are favored by relatively flat terrain. Tourists flock to the country, to enjoy a land where spectacular volcanic peaks rise above idyllic beaches and warm ocean waters. Indonesia may be best known, however, for its treacherous terrain.

Treacherous Terrain

Geologically, most of the country lies astride one of the world's most active and hazardous zones—the Pacific Ring of Fire—and several branching tectonic plates (huge moving masses of the Earth's crust). In fact, more than 100 of Indonesia's 400 volcanoes remain active; 70 eruptions have occurred since 1600. Of the 23 volcanic disasters since A.D. 1000, in which more than 1,000 lives were lost, 9 have been in Indonesia. Java and Bali are two of the world's most volcanically active spots. There, and in many other regions of the country, people are quite familiar with the belching ash, acrid smoke, fiery clouds of hot volcanic gases, and glowing lava attributable to the angry natural forces. In addition, each year the country experiences an average 500 earthquakes, some of which are severe. Both volcanic explosions and seismic activity can cause tsunamis, great sea waves, that can devastate coastal regions. (For additional information on the Ring of Fire, enter "Pacific ring of fire" into any search engine and browse the sites listed.)

Two of History's Greatest Eruptions

Indonesia experienced two of the most severe volcanic eruptions in recorded human history. In April 1815, on Sumbawa

Indonesia is the world's largest archipelago, and its chain of islands stretches 3,000 miles (4,800 kilometers) from east to west. Though it has a tropical location, in addition to pristine beaches it is also home to snow-capped mountain ranges and potentially destructive volcanoes.

Island, located 200 miles (320 kilometers) east of Java, the Tambora Volcano erupted violently. An estimated 12,000 people lost their lives immediately. Another 80,000 died of starvation because so many food-producing fields were buried under volcanic ash. Beyond the islands, effects of the explosion were also experienced throughout much of the globe. Ash from the eruption was thrust high into the atmosphere. There, it blocked some of the sun's rays for several years, causing a decrease in the Earth's temperatures. In fact, 1816 was called "the year without a summer," because the atmospheric effects of the 1815 eruption lowered global temperatures by over 5° F (about 3° C). This cooling effect was much greater than that resulting from the 1991 eruption of Mount Pinatubo in the Philippines, which lowered temperatures by a little less than 1° F (.5° C).

Although perhaps not as catastrophic as the eruption of Tambora, the 1883 explosion of Krakatoa (Krakatau) is better documented and more widely known. Geographically, the volcanic island was misplaced in the title of a popular 1969 film, "Krakatoa, East of Java." It actually lies to the west of Java in the Sunda Strait. On August 26, 1883, residents of Rodriguez Island, located in the western Indian Ocean off the coast of Madagascar, could hear a low and distant rumble, sounding much like thunder at a distance. Little did they know that it was the sound of a catastrophic volcanic explosion that occurred nearly 3,000 miles (4,800 kilometers) distant, in Indonesia!

A fascinating book by Thomas Simkin and Richard Fiske documents the Krakatoa eruption, which was one of the most violent events of recent history. The explosion destroyed two-thirds of the island, blowing rock debris up to 17 miles (23 kilometers) into the atmosphere. Where land once had been, the water was up to 1,000 feet (300 meters) deep. The explosion caused a giant tsunami, with waves reaching 120 feet (40 meters) in height. Along low-lying coastal areas, the wall of water roared up to 6 miles (9.5 kilometers) inland. By the time the water had receded, 165 coastal villages had been destroyed and between 36,000 and 37,000 lives had been lost. Atmospheric ash and gases from the eruption caused vivid sunrises and sunsets for three years. Finally, the global temperature dropped about 2° F (1.2° C) for a year and did not return to normal for five years.

Other Volcanic Hazards

Volcanoes pose other hazards: They can collapse, creating calderas (craters, such as Crater Lake National Park in Oregon). Caldera-formed lakes are often scenic, but the collapse of a mountain can be deadly if the area is populated. Lakes can also form in volcanic craters. In 1583, magma rose beneath the crater of Kelut, one of the country's most active volcanoes, located on the island of Java. The resulting lahars (boiling mudflows) churned down adjacent valleys, killing an estimated 10,000 people. Another 5,000 lives were lost in similar events in 1919.

WEATHER AND CLIMATE

Imagine living in a place where every day of your life the weather was nearly the same. Welcome to the tropics, where weather (daily atmospheric conditions) and climate (long-range average of weather) are incredibly monotonous! Both temperatures and precipitation are strongly influenced by two dominant controls: a tropical location and marine or maritime influence. Both conditions contribute to a constantly warm and moist condition, but one lacking marked seasonal variations.

Differences in temperature between night and day range from 70° to 90° F (25° to 32° C), a span 10 times greater than those marking seasonal changes. Seasonal temperatures, on the other hand, vary by only a degree or two. Throughout much of Indonesia, the range from coldest months to warmest months is from 78° to 80° F (25.5° to 26.7° C). This is why "nighttime" is often called "the winter of the tropics." Most of the country has never come near experiencing frost, although high mountaintops are snow-clad throughout the year, offering an amazing contrast in climatic conditions.

Monotony prevails in precipitation, as well. In fact, nearly all of Indonesia receives more than 80 inches (200 centimeters) of rainfall a year, with some areas receiving up to a drenching 400 inches (1,000 centimeters) a year—ranking them among the world's wettest locations. Bogor, a city located in the mountains south of Jakarta on the island of Java, holds the world record for the average number of days with thunderstorms—a whopping 322! Indonesian mountain slopes do experience a slight seasonal variation in rainfall because of monsoons (periodic winds). Because the country's islands lie in both the northern and southern hemispheres, patterns are quite complex. Some parts of the country, and even some individual islands because of the influence of their mountains (rain on the windward side, dry on the leeward side), show marked differences in their wetter/drier seasons. No part of the country, however, is subject to prolonged drought.

PLANT AND ANIMAL LIFE

Tropical environments offer the Earth's greatest abundance of plant and animal life. Indonesia is no exception. It is rich in both floral and faunal species. Most of the region was covered initially by a vast and lush tropical rain forest. At present, hundreds of thousands of species can occupy an area of a square mile, including a hundred or more species of trees alone. Towering emergent trees often reach 200 feet (60 meters) skyward, in search of sunlight. Below the dense canopy formed by their crowns lives a variety of shade-loving tree species and a vast array of shrubs, bamboo, canes, mosses, and ferns. The forest floor, itself, is often clear and open, because it is difficult for many plants to thrive without sunlight.

A Biogeographical Mystery Solved

During the middle of the nineteenth century, English scientist Alfred R. Wallace made a fascinating discovery. He recognized that striking differences existed between the plant and animal life on two sides of a "line" running through Indonesia—a line recognized today as "Wallace's Line." To the west of the line, which includes the islands of Bali, Java, and Borneo, Wallace noted the dominance of Asiatic life forms. To the east, which includes Lombok, Celebes, Halmahera, and other islands, he found that Australian life forms dominated.

These differences, he realized, indicated a separation between the two continental masses at some early date in history. They also suggested that lands to the west of the line had, at one time, been a part of the Asian mainland and lands to the east had been connected to Australia. Scientists now recognize that this did, indeed, happen. During the Ice Age, the global sea level dropped several hundred feet. Much of what is now seafloor to the west of Wallace's Line was then exposed as dry land, across which animals and plant freely spread. This movement stopped for most species, however, when open water was reached.

Wallace's initial observations were based on sharp differences he observed in bird species living on islands either side of the line. Soon, however, other types of animal and plant life were recognized as being vastly different, as well. To the west were Asian plants and animals, including the tiger, rhinoceros, deer, buffalo, and wild pig. To the east, the only native mammals are marsupials (pouched animals) and monotremes, both of which lay eggs and suckle their young; these are also common to Australia. Plant life also corresponds to the line, with Asian species dominating to the west and Australian species, particularly eucalyptus, dominating to the east.

Animals You Want to Meet Only in a Zoo

Indonesia boasts a rich variety of animal life—including both fresh- and saltwater fish and other aquatic organisms, hundreds of bird species, and, of course, countless types of insects. The country also provides a habitat for some life forms that you might want to visit only in a zoo.

Some 450 species of snakes, including a number that are highly venomous, make their home in Indonesia. The latter include the deadly coral snake, banded krait, king cobra, and several varieties of vipers. Warm tropical waters surrounding the islands teem with millions of sea snakes, which, although among the world's most toxic, fortunately are quite non-aggressive. Indonesia is also home to the world's largest snake, the reticulated python. For decades, the world record for length was held by a 33-foot (10-meter) long python found on the island of Sulawesi in the early 1900s. In December 2003, however, an animal exhibition park in Kendal, a town located in central Java, made a shocking announcement. It claimed to have on display a huge python measuring 48 feet, 7 inches (14.83 meters) long and 2.8 feet (.85 meters) in diameter, and weighing 984 pounds (446 kilograms) [Reuters, 30 December 2003].

Indonesia also is home to the world's largest—and perhaps most dangerous—lizard, the Komodo dragon. These astounding reptiles are found on a small number of islands, including Komodo and Flores. They grow up to 10 feet (3 meters) in length and can weigh more than 200 pounds (90 kilograms). Komodos are good swimmers, can move short distances as fast as a human, and have keen eyesight and sharp claws. Deadly bacteria in the dragons' mouths cause infections, from which the prey slowly dies after being bitten. Today, only an estimated 2,500 to 5,000 Komodos remain, and their habitat becomes increasingly endangered by human activity.

ENVIRONMENTAL ISSUES

Indonesia, like many less developed countries (LDCs), suffers from a number of environmental problems. Air pollution, water pollution, sewage, garbage, and industrial wastes make

Indonesia is home to a variety of animal species, including the Komodo dragon, seen here. Found on a few islands throughout the country, the Komodo dragon is the world's largest lizard, sometimes growing 10 feet long and weighing over 200 pounds. The dragons are quick and deadly, producing bacteria in their mouths that slowly kills prey that have been bitten.

many urban environments quite toxic. The widespread practice of burning forests to clear the land for agricultural use often causes severe seasonal air pollution. Much of the country's wildlife is threatened by loss of natural habitat as human populations swell and require more land for settlement.

Deforestation is perhaps the leading environmental problem in Indonesia today. The country's forests are rich in plant species prized for their high commercial value. Teak, ebony, mahogany, sandalwood, and several other species attract illegal logging, which is rampant on many islands. For a fee, corrupt government and military officials will often turn their backs on illegal logging operations. Much of the country is being rapidly stripped of trees. Not only does this remove an important economic resource, but it also exposes hillsides to severe erosion and the loss of valuable soil. Natural forests also are being replaced with rubber, coconut, and oil palm plantations, and kapok (*Ceiba* genus) trees.

Many aspects of Indonesian society and culture have been influenced by patterns of human cultural adaptation to the country's scattered territory, an often violent nature, and tropical conditions. The historical aspects of this relationship, and other topics, will be discussed in the following chapter.

CHAPTER

3

Indonesia Through Time

How could a humanlike creature that lived hundreds of thousands of years ago in Indonesia cause a huge controversy in late nineteenth- and early twentieth-century Europe? This is what happened in 1891, after a skullcap was found by a Dutch physician, Eugene Dubois. Dubois had gone to the island of Java to look for evidence that would support Charles Darwin's theory of evolution. He believed that a "missing link" between humans and apes could be found. Furthermore, he was convinced that evidence bridging the gap between apes and humans would most likely be found in a tropical setting. Based on this belief, Dubois concentrated his search on the island of Java. Here, he found a skullcap, and, only 50 feet (15 meters) away, a femur bone. The following year, what appeared to be human teeth also were discovered. Believing this to be the missing link, Dubois touted his discovery in Europe. His find set the scientific and religious communities into a tizzy of controversy. The outcry eventually became so great that Dubois buried the skull and bones under his home, where they stayed for nearly 30 years.

Dubois's finding has come to be called *Pithecanthropus erectus,* or Java man. Java man lived in the Paleolithic age and provides an ancient link to human roots in Indonesia. The controversy over whether Java man is human, ape, or a "link" between them, still rages on the Internet as evolutionists and creationists debate the meaning of this discovery. Disagreement also exists concerning Java man's antiquity. Estimates range from 500,000 years to as far back as 1.7 million years. Geographically, our only concern is that humanlike creatures of some type have been in what is now Indonesia for a very long time.

Since Java man, Indonesia's history has been like a stew. First there are the ingredients in the stew that are indigenous (native) to the islands. These homegrown ingredients are represented by the islands' great variety of different native cultures. These cultures developed in relative isolation from each other because of the difficulty of transportation and communication in ancient times. Add to this basic stew the ingredients of various colonial masters such as the Portuguese, British, and Dutch. Each of these colonial influences added more flavorings that have created the complex mixture that is evident in Indonesia's culture and history. Then add additional ingredients of religion and a period of Japanese occupation. By the time the stew is completed, either you have a wonderful broth, or too many cooks have been in the kitchen and created a mess. Both scenarios are legacies of Indonesia. This chapter examines the historical and cultural roots of Indonesia and the cultural elements that have been added over time.

PREHISTORY AND INDIAN KINGDOMS

Indonesia's prehistory indicates that the earliest "modern" humans arrived in the archipelago about 40,000 years ago. They most likely came from the Philippines, Thailand, Burma, or other areas on the Malay Peninsula. This migration took place during the Ice Age, when freezing long-term temperatures exposed land bridges. Later, settlers arrived from the Indian subcontinent and other areas of Asia. They brought agricultural techniques such as the growing of rice; they also brought pottery and stone tools, evident from about 2500 B.C. to 500 B.C. These immigrants intermarried with people from indigenous cultures, as the cultural stew of Indonesia started to take early form. Between 500 B.C. and A.D. 500, contacts with mainland Asia developed further. Of particular importance, metal tools and domesticated animals began to arrive.

Settlers from India came in greater numbers during the first seven centuries after the birth of Christ. They brought with them many of their customs, the Sanskrit language, and both Hinduism and Buddhism. The indigenous people on the islands retained many local cultural traditions but added new elements picked up from the Indian immigrants. Major tenets of Hinduism and Buddhism were absorbed, but Indonesians basically rejected India's notion of a caste system.

Small Indian kingdoms developed in many locations around the islands during the first millennium A.D. The greatest of these was the Srivijaya Kingdom, located on eastern Sumatra. By the seventh century, Srivijaya, a Buddhist kingdom, had developed elaborate trade and political relationships with locales in India, China, and elsewhere in the archipelago. The kingdom was able to extend its influence to other islands and eventually developed strongholds on Java and even the Malay Peninsula. Srivijaya had a strong naval fleet and dominated the waters around the kingdom until the thirteenth century. This allowed Srivijaya to control the region's lucrative spice trade for nearly 400 years. Srivijaya also became an important intellectual and religious center, with hundreds of Buddhist monks teaching and studying in the capital city of Palembang. Palembang was also the major port city in Srivijaya. Some historical reports claim that occasionally as many as 1,000 ships were docked in the city's

Settlers from India came to Indonesia in the first millennium A.D., bringing their cultural and religious traditions. This Buddhist-Hindu temple in central Java was built in the ninth century by Srivijaya, an Indian Buddhist kingdom. It is one of the most important Buddhist-Hindu sites in the world.

harbor. By the fourteenth century, Srivijaya fell into a steep decline and split into eight smaller kingdoms. The most important of these was Malayu, which also became a sea power, but not on the same scale that Srivijaya had been earlier. Today, little remains of the Srivijaya kingdom, although Palembang has become an important city with a population of about 1,500,000.

ISLAMIC KINGDOMS

Asian immigrants and Hinduism were but the first of many cultural waves to splash upon Indonesia's shores. The next breaker to reach the archipelago was Islam. Arabs had been trading with Indonesians since the fourth century. By the thirteenth century, when Srivijaya was still a power, the Indian Ocean served as a major sea bridge between the cultures that bordered it. The prophet Muhammad founded Islam on the Arabian Peninsula in the seventh century; it spread quickly and found its way to Indonesia by way of the Arab merchants who brought their new religion along with the items they wished to trade. Many coastal Hindu kingdoms, such as Demak on Java, and their kings, were converted to Islam. At the same time, many inland areas of the islands, such as Java, were less accessible to the Muslim traders. People there often held onto their Hindu and other traditional beliefs. Other areas, such as Ambon, fiercely resisted the Islamic faith in favor of Christianity, which took hold later.

Islam did not arrive as an invasion. Rather, it was gradually infused into the many local cultures that resided along the archipelago's coastlines. Coastal areas on Java became strongly Muslim. Kingdoms there adopted the new faith and combined it with traditional beliefs, as Indonesians have always done. When Marco Polo stopped on Sumatra on his way home in 1292, he reported visiting an Islamic coastal community named Perlak, which was then surrounded by non-Islamic communities. After Islam had developed strongholds, some of these kingdoms attacked their Hindu-Buddhist neighbors in attempts to convert the people living there. Small sultanates continued to develop along the coastlines as Islam took deep root in most of the archipelago. Some exceptions remained in Bali, West Papua, and Flores, where the localized Hindu-Buddhism traditions prevailed.

By the fourteenth century, belief in Islam was strong on the islands. This religion brought the people a number of new ideas in the area of commerce and trade. It also served as a protective wall between local peoples and the arriving Europeans. Arabic language can be found on artifacts dating back to this time period, suggesting that Islam was exerting a powerful cultural impact upon art, literature, science, and politics. The process of converting Indonesia to Islam took hundreds of years, however, extending well into the seventeenth century. After the arrival of Europeans, Christians and Muslims began to vie for spiritual followers across Indonesia.

ARRIVAL OF EUROPEAN INFLUENCES

Although Europeans visited the islands earlier, the first significant European contact was by the Portuguese in 1511. In the early sixteenth century, the Portuguese were a mighty sea power. Mostly as a result of the efforts of Prince Henry the Navigator in the fifteenth century, the Portuguese had excellent ships and superior navigating technology. They became deeply involved in the lucrative spice trade and were charged by the pope with the responsibility of converting Asians to Christianity.

With their superior technology, the Portuguese sought to wrest control of important shipping lanes in the Indian Ocean from Arab traders. Under Admiral Albuquerque, Portuguese naval forces captured the important Islamic port of Malacca on the Malay Peninsula. This allowed them to control the sea-lanes through the vital Strait of Malacca. Looking at a map shows how key the narrow Malacca waters are to controlling traffic through the strait. Sometimes the Portuguese even resorted to piracy in their efforts to monopolize the valuable trade in pepper, nutmeg, cloves, mace, ginger, and cinnamon. Some of these spices were more valuable than gold in the markets of European cities. Black pepper, for example, increased in value by 10,000 percent by the time it was sold in Europe.

Islamic sultanates and traders often worked to resist the Portuguese. In 1570 the Portuguese assassinated the Sultan of Ternate in an effort to have a more favorable successor. This effort failed miserably. The local population revolted against the Portuguese in 1574 and forced them off the island. Since Portugal was a small European power, the revolt on Ternate, a small island off of the coast of Halmahera, seemed to start a chain reaction that ultimately pushed the Portuguese out of the archipelago. By 1641 they had lost the important port of Malacca. The last remnant of Portuguese dominion fell in 1974, when they gave up their East Timor colony.

What flavorings did the Portuguese add to the stew we call Indonesia? They obviously failed in their papal assignment to establish Christianity in the region, but some influences were left behind. For example, the Portuguese language can be found in hundreds of words and place names in Indonesia. Influences in music are also evident, as is tobacco, which was introduced by the Portuguese. Another lasting impact of Portuguese influence was the superior shipbuilding techniques they introduced.

The English also landed on the shores of the archipelago. Their time of arrival paralleled that of the first Dutch. Both countries were maritime superpowers that began competing to control the lucrative spice trade as Portuguese influence in the region waned. Even though the two European powers had a treaty that compelled them to cooperate in trade activities, in reality their relationship was rife with competition and animosity.

The high point for the British in Indonesia was when they briefly ruled the area known as Java and Dependencies, starting in 1811. British rule received a boost as Dutch attention shifted toward home, where the French, under Napoleon Bonaparte, occupied the Netherlands. Thomas Stamford Raffles was appointed lieutenant governor general of the region. Raffles is perhaps best recognized as the founder of Singapore, in 1819. Under Raffles, a number of important changes took place. He initiated reforms in slavery, introduced limited self-government, and encouraged land reform and crop selection by locals. British rule only lasted until 1816, however. When Napoleon was defeated, the British promptly returned rule to the Dutch, in accordance with an agreement between the two countries.

DUTCH RULE IN INDONESIA

The largest wave to wash over the archipelago—a virtual cultural tsunami—was the influence of the Dutch, who served as colonial master of the islands for hundreds of years. Many flavors were added to Indonesia's cultural stew during the time that Holland served as a colonial master.

In 1596, four Dutch ships arrived at Banten, a port in western Java. Their mission was to obtain spices and return home to Holland with the treasured cargo they obtained in the Spice Islands. The Dutch quickly realized the extremely high value of Indonesia's precious spices. More Dutch ships quickly returned to exploit the potential of the Spice Islands. In 1602, the Dutch chartered the Dutch East India Company [Vereenigde Oost-Indische Compagnie (VOC)] to control and monopolize trade in the archipelago. They quickly invaded Portuguese strongholds on the Banda Islands (part of the Moluccas) and forced them out. Dutch warships began to appear in Indonesian waters, as they took steps to annex the islands and turn them into a Dutch colony under VOC rule.

The VOC had immense powers, stipulated by a charter granted by Holland's parliament. It was a stock-held company run by a 17-member board called the Seventeen Gentlemen. Powers granted in the charter included the right to convene military troops, to build forts, to forge treaties with local leaders, to conduct trade, and to operate the courts for the region. The colonial hand of the VOC reached deep into daily life in Indonesia, a factor that often caused local resentment. The VOC worked to isolate the Dutch East Indies, as they were now called, from the outside world. The Dutch would sink foreign ships as their grip tightened on the archipelago, so that their valuable trade monopoly in spices would be protected.

Jan Pieterszoon Coen was a two-time Governor General of the VOC and a ruthless manager. To raise the price of certain spices, such as nutmeg and cloves, he ordered production to be cut back. This was accomplished by destroying large quantities of the crops, an action that devastated the local economy in the affected areas. Strong local resentment built up in some areas, and Sultan Agung attacked Coen's fort at Batavia in 1629. Batavia is now called Jakarta. The VOC defeated Agung's onslaught, and the Dutch razed the city as punishment. However, in the failed attempt to take the city, Coen was killed. Other rebellions led by Sultan Hasanuddin, Prince Trunojoyo, and Jakarta's Chinese population tried to push the Dutch company and its forces out of the islands, but all of these efforts ultimately failed and resulted in further Dutch punishment and harsh rule. Local people who resisted the Dutch intrusion into their crop production were subject to various atrocities and even death. The Chinese rebellion in Jakarta alone resulted in the deaths of more than 10,000 people. Finally, because of corruption and poor management, the VOC went bankrupt in 1799. What came next was little improvement, as the Dutch government took over.

"Culture" or "Cultivation System" sounds like an innocuous and friendly name for a program. However, the program itself tightened the grasp of the harsh Dutch hand on Indonesia. This system forced production of agricultural products that were desired by European markets. This meant that coffee, sugar, pepper, indigo, and cotton were raised in plantations that used locals as slave-like laborers. The working conditions and living conditions were extremely poor for local people. In fact, thousands of workers actually starved to death in the major rice-producing area of Cerebon in 1849–1850. This system of production resulted in huge cash profits for the Dutch and obedient local residents. By 1938, there were nearly 2,400 of these plantations. The budget of the Dutch government was actually being balanced by the labor of poor Indonesian workers. In addition, the Dutch believed that they were racially superior to the local people, whom they regarded as half devil and half child. For example, a sign at a Dutch swimming pool even said "No Natives or Dogs."

Local resistance to the Dutch waned after the Java War (1825–1830). Prince Pangeran Diponegoro of Mataram led the doomed struggle for freedom from the Dutch. The Prince was an aristocrat who felt wronged when the Dutch tried to build a road across his land. In guerrilla fighting over five years, more than 200,000 Javanese were killed or died of related causes. Even in failure, Diponegoro is considered a hero in Indonesia today.

After the Java War, the Dutch worked to add to the Dutch East Indies by slowly taking formerly independent areas in the archipelago. This usually resulted in more violence, but regions slowly fell to the Dutch during the century following the Java War. Kalimantan was added in 1860, Banjarmasin in 1864, Madura in 1887, Lombok in 1894, Sulawesi in 1905, Bali in 1906, Aceh in 1908, and finally West Papua in 1920. By the end of the Dutch consolidation, the Dutch East Indies represented an area roughly the same as that of today's Indonesia.

Eventually, Dutch rule changed from the harsh Culture or Cultivation System to the Liberal System, which operated from 1870 to 1900. This system was intended by the Dutch to be more humane. It sought to end ruthless exploitation of the native people and correct injustices. A movement started in the Dutch Parliament during a time now called the "ethical period" sought to improve human conditions in the colony. This effort included improving flood control systems, agriculture, education, health care, irrigation programs, and a program designed to address the issue of overpopulation on Java. The Dutch abolished slavery in the colony in advance of the Liberal System. But they introduced another program that also posed problems. This Liberal System program forced local people to move to outlying islands in an effort to reduce the very dense population of Java and to populate, hence develop, underutilized areas. Tragically, many of the settlers were subject to extreme ethnic persecution in the new areas. Ultimately, the program, and the entire Liberal System, proved to be largely unsuccessful.

The indigenous population did not just sit back and watch, however, while the Dutch continued to rule the archipelago. Many of the native peoples and local kingdoms resisted fiercely, but the Dutch prevailed. When local resistance failed, some Indonesians saw the need for a broader and better-organized effort against the Dutch. Influenced by early twentieth-century nationalistic movements in other countries around the world, local people in the archipelago saw that colonial powers could be pushed out with a unified effort.

Efforts to organize a Dutch resistance started with the founding of Budi Utomo. This organization was originally established by intellectuals for educational endeavors and was primarily connected to training native doctors. Budi Utomo was founded by Soetomo in 1908; the name of the group means "noble conduct." Displeased with Indonesia's undesirable colonial status, the organization soon became engaged in politics. This group is considered to be Indonesia's first modern political organization. Over the next few years (1910–1920), additional groups such as the Indies Party, Sarekat Islam (Islamic Union), Communist Association of the Indies, and Muhammadiyah (Followers of Muhammad) were formed to resist Dutch colonial rule. The first group believed to have used the term *Indonesia* (contraction of *India,* "cultural influence" and *island*) was a political student group who met in the 1920s and called themselves the Indonesian Alliance of Students.

Budi Utomo chose to use Malay as its official language instead of Javanese. This was a significant decision. Later, the language became adopted as the official national tongue that is now called Bahasa Indonesia. The decision to use the Malay language in the anti-Dutch colonial movement served as a linguistic unifying factor in developing the nation of Indonesia.

In the late 1920s, Sukarno started his rise to power. From humble beginnings, he rose to have a national political reputation because of his work and association with the

Indies Party and Sarekat Islam. He, along with others, established the Indonesian Nationalist Party (PNI) in 1927. This party promoted noncooperation with the Dutch and the strategic objective of achieving independence. Sukarno was a broad thinker and tied together three important considerations in his program for nationhood: first was the importance of traditional Indonesian cultures, second was Islam, and the third was ideas drawn from Karl Marx. His outspoken philosophy caused him to be imprisoned by the Dutch, who had long abandoned their Liberal Policy. They were deeply concerned about the rising fury and cry for independence, which made the Dutch more skeptical and intolerant of leaders like Sukarno. Dutch persecution of parties seeking independence caused many of the groups to dissolve in the 1930s. The idea of an independent Indonesia promoted by Sukarno and other political leaders would not die.

For centuries, the waves of foreign colonial powers washed upon the shores of Indonesia. The Dutch had been a tsunami that overran the country. For periods of time, they offered unenlightened rule that subjugated the native people of the archipelago to slavery, poverty, forced migration, difficult living conditions, and unequal treatment. Now the seeds of the country of Indonesia were taking root in the nationalistic movements that started during the first decades of the twentieth century. Would further waves of foreign domination wash upon the shores of Indonesia before independence? How would the Dutch be pushed out? These questions are answered in the following chapter.

CHAPTER

4

Independence and Nationhood

Indonesia's journey to independence did not happen easily or predictably. As the storm clouds of World War II gathered in the Pacific, Dutch colonial rule persisted. Japan began its quest to become a colonial power with the 1931 invasion of Manchuria, an event that served as a prelude to World War II. Island hopping across the Pacific, Japanese forces gained control of the Dutch East Indies in May 1942, and they occupied much of Southeast Asia by 1943.

ERA OF JAPANESE OCCUPATION

For Japan, taking the Dutch East Indies was not difficult. The Netherlands (Holland) had been overrun by the Germans in 1940 and was under their occupation in 1942. Further, Japan had promoted its arrival as signaling the liberation of the islands from the Dutch; consequently, they were enthusiastically received. Once they gained a foothold in Indonesia, however, the Japanese found a

Japanese forces had gained controlled of the Dutch East Indies and much of Southeast Asia by the latter stages of World War II. Here, Japanese warships enter Camranh Bay in 1941.

wealth of badly needed resources, including rubber, tin, and petroleum. The islands were divided into three administrative units, and the Japanese proceeded to occupy the islands and to begin extracting natural resources. To placate the locals, Japanese military leaders allowed some political activity by Sukarno and others. Thus, for their own purposes, Japan fueled the quest for independence in their efforts to control the population. Sukarno and other Indonesian independence leaders cooperated with the Japanese commanders. They saw the occupation as a possible path toward removing the Dutch and securing independence.

The impact of Japan's three-year occupation was mixed. More than 4 million Indonesians were subject to cruel, forced labor to benefit Japanese defense or economic development. The Japanese pushed many men into the military as conscripts. Thousands of females were forced to become "comfort women"—they were enslaved in sex camps that served the Japanese military. (Much later, Japan worked to compensate those who had been forced to be comfort women during the war. Cash was offered to the women for the horrible fate they had suffered during the occupation.)

For the Dutch, it was also a difficult time; the Japanese imprisoned nearly 200,000 Dutch in internment camps during the war.

Despite the horrors associated with the Japanese occupation, the independence movement made huge steps forward. Through Sukarno and Muhammad Hatta (another nationalist leader appearing to cooperate with the Japanese), the Japanese appeared to encourage political activity, because it reduced resistance to the occupation. During the war, the Dutch language was discarded by the nationalists in favor of Bahasa Indonesia and Japanese. Indonesians now viewed the Dutch as weak colonial masters who easily could be thrown out after the war, much like their language had been cast aside during the occupation.

INDONESIA ACHIEVES INDEPENDENCE

Japan surrendered to the Allies on August 15, 1945. Prepared for this outcome, Indonesians led by Hatta and Sukarno immediately seized the military weapons of the Japanese for possible future use. Sukarno and Hatta then declared Indonesian independence two days later, on August 17. A constitution had already been written, during the spring of 1945. This document incorporated the five basic ideas of Sukarno's Pancasila, the state ideology that established a foundation for independence. Pancasila's core beliefs were in God, humanitarianism, national unity, democracy, and social justice. The new Constitution was announced on August 18, 1945; Sukarno assumed the presidency and Hatta became vice-president shortly afterward. Events were moving fast, and the Dutch were not happy about these circumstances.

The Dutch despised Sukarno and Hatta and labeled them as Japanese collaborators. They were limited, however, in what they could do to reoccupy and control the colony of the Dutch East Indies. Their primary attention continued to focus on the lingering effects of the war at home. Fearing restoration of Dutch colonial power by the Allies, the new government quickly moved to consolidate power and establish control over the new and fragile country. The weakened Dutch initially tried to form some type of commonwealth relationship with Indonesia. But fighting soon broke out in various locations throughout the islands. Communists rebelled in East Java, where earlier in the postwar period the British, representing the Allies, had fought with local insurgents, resulting in thousands of deaths in what was the largest battle for independence. Dutch troops tried to push the nationalists out of Java and Sumatra. Islamic groups attempted to form their own country on Western Java. The communist threat appeared to be growing.

The United States and other Allies were increasingly concerned over the growing communist threat in Indonesia and elsewhere in the world. The "red menace" was a major concern of the United States and its Allies after World War II. The United States viewed the Indonesian nationalists as anti-communist, however, because they had fought against various communist insurrections in the archipelago. Consequently, the United States and others applied pressure on the Dutch to recognize Indonesia's independence. Finally, on December 27, 1949, the Netherlands transferred all powers to the Republic of Indonesia, with Jakarta designated as the capital.

SUKARNO'S TURBULENT ERA

Independence did not resolve all the major issues facing Indonesia's new government. There were still separatist movements in various parts of the country. The Dutch still held West New Guinea and the Portuguese East Timor. Further, Indonesians continued to be widely divided on the basis of religious, regional, class, political, and ethnic differences. The economy was also in disarray, as the Dutch had been thrown out and inflation was becoming a big problem. Corruption, injustice, and inefficiency also started appearing in the Sukarno government. Nearly 60 political parties added additional chaos and confusion to the political landscape. A country composed of thousands of islands and many diverse beliefs made ruling extremely difficult for the somewhat inept new government.

Although a democracy after independence, Indonesia had little experience with democratic processes and principles. The examples left by Japan and the Europeans were more authoritarian in nature and did not prepare the country for running itself as a democracy. Thus, by the mid-1950s the military started assuming a larger political role. However, the military was also divided. After Sukarno and his party were narrowly kept in office in a general election in 1955, Sukarno began shepherding the nation toward what he called "Guided Democracy." In actuality, however, this movement strengthened Sukarno and his presidency and moved Indonesia toward a more authoritarian government.

Some people, both within and outside the government, recognized Sukarno's authoritarian moves and protested his policy of Guided Democracy. Vice-president Hatta even resigned in 1956 because of Sukarno's growing authoritarian rule. In 1957, Sukarno declared martial law, as he and various military factions sought to preserve national unity. Sukarno strategically held together a fragile coalition that favored the military and the communists. In 1959, Sukarno dissolved the House of Representatives and a year later set up his own new House of People's Representatives. In 1963, he declared himself president for life.

At the same time, the United States and other countries were growing increasingly suspicious of Sukarno. This distrust

Achmad Sukarno rose to power in the 1920s due to his work to achieve Indonesian independence from the Dutch. Japanese occupation of Indonesia ended with the end of World War II in 1945, and Sukarno assumed presidency. Here, in 1950, Sukarno speaks to a crowd protesting his leadership.

grew stronger when he began drawing closer to communism. Sukarno openly sought the advice of communist leaders and drew his country nearer to the Soviet Union and Communist China. In addition, he repudiated the United States and withdrew Indonesia from the United Nations in 1965.

Holding together his country's fragile politics was a balancing act that Sukarno had conducted for decades. Playing the military, religious groups, the communists, and other interests against each other was a dangerous juggling act that Sukarno performed to survive politically in his efforts to hold the country together. Despite these efforts, however, the country was falling apart politically, socially, and economically. Guided Democracy had failed. A coup attempt by military communist party sympathizers took place on September 30, 1965. The group said they were conducting a preemptive coup to prevent a takeover by Indonesian military leaders under the influence of the U.S. CIA. Five generals were killed in the failed coup attempt. Even though the coup failed, this attempt served as an important initial step toward removing Sukarno. The coup triggered strong and violent protests against Sukarno's government. He had 300,000 Chinese and communist sympathizers killed in the bloody aftermath of the failed coup. With this action, Sukarno lost his communist allies and was eventually replaced by the military. He was stripped of his title of President for Life and thrown out of office, to be followed by General Suharto in 1966. Sukarno lived under isolated house arrest until his death in 1970.

THE SUHARTO ERA

Many observers believe that the U.S. CIA and British MI6 intelligence agency played a major role in the overthrow of Sukarno. Both countries viewed Sukarno as a potential communist threat during the Cold War. The government that followed Sukarno was more pro-West. Under the new Suharto leadership, the country reestablished relations with the United Nations in 1967 and developed relations with other countries. On the economic front, in 1967 Indonesia became one of five founding members of the Association of Southeast Asian Nations (ASEAN).

Suharto's governance was also marked by authoritarian rule. He called his plan the New Order. Under the New Order, he worked to suppress and persecute the communists, but he exhibited patience with other political parties. The military played a strong role in his government and even operated many business enterprises. In part because of strong military support, stability increased during Suharto's leadership; this also brought back the principles contained in Pancasila. The following were key elements of Suharto's New Order:

1. Complete the restoration of order and security and establish political stability.
2. Carry out economic rehabilitation.
3. Prepare a plan for national development and carry out the plan with an emphasis on economic development.
4. Return to the United Nations.
5. Pursue an active and independent foreign policy.
6. Establish Indonesia's economic credibility overseas.
7. Hold general elections once every five years.

The economy did improve under Suharto. Oil had long been an important economic resource for the country, and the price of petroleum soared during the 1973 international oil crisis. As a member of OPEC (Organization of the Petroleum Exporting Countries), Indonesia's price for oil increased with the unity offered by the producers' organization. In addition to an improving economy, East Timor reluctantly became part of Indonesia in the mid-1970s. The economic recovery lasted until the 1997, when excess foreign debt and inflation caused the Rupiah, Indonesia's currency, to plummet. The economy went into a yearlong downward spiral.

Suharto was elected president seven times, and served in this office from 1967 to 1998. One of the great achievements of his presidency was the increased consensus within the diverse country. The economy also greatly expanded under Suharto, although he and his family also became extremely wealthy. Because of the wealth amassed by Suharto and his children and resulting allegations of major government corruption, he was forced to resign in 1998 amid widespread violence and bloody riots. Today, he is one of the richest people in the world. Efforts are being made, however, to hold him legally accountable for alleged ill-gotten funds.

INDONESIA'S GOVERNMENT ENTERS THE TWENTY-FIRST CENTURY

Abdurrahman 'Gus Dur' Wahid was elected President in 1999 in Indonesia's first democratic election after Suharto and served as a brief caretaker president. The legacy of Sukarno's political party, called Indonesia Democracy Party-Struggle, was led by his daughter Megawati Sukarnoputri. This party actually came in first in the elections, but Sukarnoputri could not build the coalition necessary to form a government. Thus she was chosen to be vice-president.

The economy began to improve at the dawn of the twenty-first century, but President Wahid found himself caught in corruption scandals that caused him to rapidly lose his popularity. Despite being found innocent of corruption, the allegations resulted in Wahid's removal from office in 2001. His vice-president, Sukarnoputri, became Indonesia's first woman president. Ethnic and religious divisions continued to exist in the early years of Sukarnoputri's presidency. Adding to her troubles was a new problem—in the form of terrorists associated with al Qaeda, who bombed a nightclub in Bali in October, 2002.

Despite the potential unifying ideals embodied in Pancasila, Indonesia remains extremely disunited by regional, ethnic, and religious divisions. East Timor became independent in 2002, and separatists in other regions, including Aceh and Irian Jaya, continue to seek independence from Indonesia. Corruption also continues to plague the government; this cancer has deep roots in the country's political culture.

History has shown that many waves have broken on Indonesia's shores over time. They have washed over Indonesia and left a variety of impressions on the human landscape. The country has withstood the potentially erosive onslaught of numerous waves of foreign cultures, to prevail over its own destiny and become independent. Indonesia, home to the world's fourth-largest population, today stands on the brink of becoming a major world power. Only the future will reveal if unity will continue to propel the country into a leading role in the world in the twenty-first century, or if Indonesia will fracture into a number of small island nations—mere pebbles, scattered across the tropical ocean.

CHAPTER

5

People and Culture

Imagine living in an isolated community—a place with limited transportation and communication technology, in a society separated almost entirely from other communities by water or harsh terrain. These are some of the factors that can severely restrict contact with neighbors, even though they may live only a few miles away. Mountains, climate, and water barriers have often contributed to the isolation of societies lacking the technology that would allow them to easily mix with their neighbors. Today, with cell phones, satellite technology, the Internet, aircraft, sophisticated ships, and other transportation systems, people have been able to overcome many of the challenges posed by their environment.

Isolation has been an important issue in the development of Indonesian culture(s). Throughout much of the islands' history, most residents have been restricted to their home island. They had little

Limited technology and harsh terrain have led to the development of many different cultures and isolated communities within Indonesia. In some areas of the country, communities are only connected through treacherous routes, like this suspension bridge over the Baliem River in Irian Jaya.

opportunity to visit neighboring islands without risking life or injury because of the dangerous ocean waters that separated communities. Others were unable or elected not to cross the difficult mountainous terrain, even when they were on the same island. These factors fostered the development of different cultures (people with different ways of living), even though the groups may be only miles apart. New technologies have helped people bridge these barriers and connect in ways not possible a few decades ago.

This chapter focuses on Indonesia's people. It includes an examination of the cultures—the religion, language, and other facets of life—that exist within the archipelago. A useful context for this examination is the ideas put forth in former President Sukarno's Pancasila, which sets forth the five overriding principles for the country. These include the belief in one God, humanitarianism, national unity, democracy, and social justice. These elements are also in the Constitution, and more important, they provide a focus for unity in this country comprising communities that have been isolated from each other for much of their history. Cultural (including linguistic and religious) and other regional divisions have served to separate many countries, including Indonesia. Sukarno's mid-twentieth-century cry for unity created a template for a modern, united Indonesia, which is even reflected in the country's holidays, in which each of the four major religions is recognized with at least one annual national holiday.

POPULATION FACTORS

Before examining major cultural aspects of Indonesia, a brief look at the country's demographics will help provide a statistical context for its people. With an estimated 234 million people, Indonesia is the world's fourth most populated country. Only China, India, and the United States have larger populations. This translates into more than 300 people per square mile (160 per square kilometer), a figure nearly four times greater than the population density in the United States. The average life expectancy for an Indonesian is nearly 69 years; women have an expectancy of 71 years and men 66. The population growth rate of Indonesia is about 1.5 percent per year. At the present rate of increase, this means that the country will have about 282 million people by 2025 and more than 315 million by 2050. The present government believes that this growth rate is too high. Only 40 percent of the population is urban, whereas 45 percent of the population is engaged in agriculture. Annual per capita income is slightly more than $700 per year, but this translates into a purchasing power worth nearly $3,000 in U.S. dollars per year.

Literacy rates are 92 percent for males and 82 percent for women, figures that reflect a social and religious bias favoring males. Women marry at an average age of 19, and 17 percent of will have 2.5 children, and 58 percent of women aged 15–64 are in the work force. This compares to 85 percent of men in the work force. Women hold only 8 percent of the seats in the national legislature and only 2 percent of the ministerial and other high government offices.

RELIGION

Islam is the dominant religion in Indonesia, and the country has the world's largest Muslim population. Eighty-eight percent of the citizens are Muslim, but the government allows other religions to exist. Included within these other government-sanctioned religions are Protestants (5 percent), Roman Catholics (3 percent), Hindus (2 percent), and Buddhists (1 percent). Only one percent of the population falls into the "Other" category of nonreligious or people holding other beliefs.

It is perhaps surprising that European religions didn't leave a larger impact on the population of the Spice Islands (the Moluccas, a part of Indonesia), or the rest of the archipelago.

This is somewhat startling after hundreds of years of colonial rule by the Portuguese, Dutch, and British. Many of these colonial powers were more interested in the islands' spices, though, than in gaining religious converts.

Islam in Indonesia

The vast majority of Indonesians, nearly 88 percent, identify themselves as Muslims. In Indonesia, however, Islam is not a monolithic religion. People in different areas of the country believe in and practice their faith in a variety of ways. In some places, Islam even has mixed with local customs and traditions to form hybrid religious practices. For example, a rural Javanese Muslim might go to the mosque to pray but later pray at the grave of a local Catholic saint. Local customs have had a similar impact on Christianity. Islam, together with some of the other major religious groups, manifests itself into the country's political parties.

The prophet Muhammad founded Islam with a belief in one God, Allah. He set forth basic tenets of the religion in Islam's most holy book, the *Koran* (*Qu'ran*), which is written in the Arabic language. Muhammad lived in Saudi Arabia in the seventh century, and the *Koran* is the compilation of writings that reflects God's revelations to Muhammad. Today, converts to Islam have now spread across much of the world. In sheer numbers, Islam is the world's fastest-growing faith.

Arab traders carried Islam to Indonesia from the twelfth to the fifteenth centuries. First contacts were in Sumatra, Kalemanta, and northern Java. Marco Polo recorded visiting a Muslim community called Perlak, when he landed in Sumatra on his way home from China. The first form of Islam in Indonesia was Sufi, an Arabic word that means wool (believers often wore woolen robes). Island converts dedicated themselves to the mystical life known as Sufi, in which believers seek a union with God. Many of these believers combined the new religious ideas with their local traditions and customs. As a result, the roots of Islam in Indonesia took on a local flavor during its early presence. Not all of the conversions to Islam were peaceful, however. Many Hindus and Buddhists fled to places like Bali to seek refuge from the wave of Islamic conversion.

Muslims believe in the five pillars of Islam, which are:

1. The testimony of faith. Allah is the one true God.
2. Prayer five times a day. Prayer is at dawn, noon, mid-afternoon, sunset, and at night.
3. Giving alms, or support to the needy.
4. Fasting from dawn to sunset during the month of Ramadan.
5. Make a pilgrimage to Mecca, Islam's most holy city, at least once during one's lifetime.

Ramadan is conducted in the ninth month of the Islamic calendar, which is a lunar calendar and different from the traditional solar calendar used in most of the world. Just before Ramadan, many Indonesian Muslims practice the tradition of visiting family gravesites to pay respect to their ancestors. This tradition often causes traffic jams in places like Java, with so many people paying their respects at the same time. Once Ramadan begins, Muslims stop eating before dawn and fast until sunset. They are also encouraged to refrain from drinking, smoking, lying, swearing, giving in to anger, and other bad habits. The time of fasting during the daylight hours is believed to be important in increasing spirituality and developing stronger self-control.

After sunset, the family may have a feast with family and friends. The end of the daily fast is announced on television and a traditional day may end with a meal bought at a Kentucky Fried Chicken franchise, as the elements of Indonesia blend with western businesses. The end of Ramadan

Islam is Indonesia's dominant religion. With 88 percent of Indonesians practicing Islam, it is the country with the largest Muslim population. Here, Muslims pray during the Eid al-Adha festival in Aceh.

is called *Eid al-fitr* in Arabic or *Lebaran* in Indonesia. This is a full-day feast that marks the beginning of a time when people visit family and friends to ask for forgiveness for wrongs they have committed over the past year.

Another important Muslim celebration is the Hajj. This occurs in the twelfth lunar month and is the time when Muslims go to Mecca to circle the *Ka'ba* (Kaaba), Islam's most holy site. *Eid al-Adha* is the celebration that marks the end of Hajj. Indonesian Muslims also celebrate Isra and Miraj, which marks the ascension of Muhammad into heaven, and Maulid

Nabi, Muhammad's birth date. (For additional information on these traditions, enter "Muslim holidays" into any search engine and browse the sites listed.)

In the late twentieth century, Islam in Indonesia began to change. More than one-half million Indonesian workers headed to Saudi Arabia in the late 1990s, and many connected with Wahhabism, a more radical, Saudi strain of Islam. This version of Islam seeks to take the religion back to its fundamental roots as contained in the Koran and to cleanse it of modern heresies and practices. Proponents of this viewpoint reject activities such as smoking, and they forbid the erection of gravestones, which they view as idols. Muslim opponents of Wahhabism contend that it is not true to the Koran and that it actually is a new fundamentalist form of Islam. Because of this disagreement, proponents of Wahhabism are also predisposed against Sufism. This extreme belief structure has caused many followers to become anti-Western. Recent terrorist acts linked to al Qaeda have also been linked to the extremist Wahhabis. The group is now developing religious schools across Java, some of which support Osama bin Laden and local groups connected to him.

It is easy to see how a conflict has been growing between the traditional Muslims and the Wahhabis, who are much more extreme and dogmatic in their beliefs, brought from Saudi Arabia. Politically, traditional Muslims are linked to the Nahdlatul Ulama (NU) political party, whereas the Wahhabis support the Muhammadiyah party. (*Muhammadiyah* translates to "the way of Mohammad.")

OTHER INDONESIAN RELIGIONS

Other monotheistic religions are also permitted in Indonesia, although their followers represent only about 12 percent of the population. The Constitution and Sukarno's Pancasila both allow for religious diversity. Christianity has been practiced in Indonesia since the 1500s.

Both Protestant and Roman Catholic influences arrived with the Portuguese and Dutch. The Dutch tried to expel the Roman Catholic Church when they defeated Portugal in 1605. They introduced their own Calvinist Dutch Reformed Church, which was the only Christian influence up until the nineteenth century. Since the Dutch church was very strict and the Dutch trading companies were focused primarily upon trade, Christianity did not advance very far during these times. In the mid-1800s, Protestant Lutherans from Germany and Catholic Jesuits from Mediterranean Europe had greater success in developing Christianity in the islands.

In the twentieth century, more missionaries arrived, and both the Catholic and Lutheran Churches made significant advances. The political overthrow of Sukarno in 1966 caused the government to view nonreligious people as possible communists, which probably helped. Thousands of suspected communists were killed when the military ruthlessly put down this rebellion. The fear people felt about such government persecution caused a further increase in the population of Christian churches in the mid-1960s. In the 1980s, Protestant churches that were becoming more localized experienced additional growth. Catholic churches, meanwhile, saw their growth stalled, possibly because the government's Muslim-controlled Department of Religious Affairs placed restrictions on missionary activities of Europeans who were working in Indonesia.

Religious violence in the islands has been on the increase in recent years—more than 6,000 people have been killed in fighting between Muslims and Christians since 1999. Islands such as Sulawesi and Maluku are fraught with religious fighting because a group called Laskar Jihad has called for a holy war against Christians. Both islands have tried to create peace agreements between Christians and Muslims, but Laskar Jihad's goal of eliminating Christianity from Indonesia stands as a formidable barrier to any hope for religious tolerance and cooperation. Religious persecution is not only by Muslims: Both Muslims and Christians have participated in a form of ethnic cleansing on the island of Ambon in attempts to remove the other religion from the island.

Buddhism has a long history in Indonesia. It is the second-oldest religion in the country, dating back to the fifth century. Buddhism peaked during the Srivijaya dynasty, which lasted from the seventh until the fourteenth century. This was a great civilization—Buddhists built colleges, monasteries, and temples that marked some of the achievements of this era. Buddhism waned after the fourteenth century, however, because Islam gained in popularity and its followers suppressed other religions. Today, only about one percent of the population claims to be Buddhist, and this number includes followers of Taoism and Confucianism because these sects are not approved and identified in the Constitution. A rebirth of Buddhism has taken shape since the mid-1950s, with monks from Thailand providing leadership. Borobudur, a temple on the island of Java that is one of the greatest religious monuments in the world is a standing tribute to Buddhism in Indonesia. The temple represents three views of the universe according to Buddhist traditions that started in India, and it has hundreds of detailed stone carvings on the superstructure. It is believed to have been completed around A.D. 825.

Hinduism is the oldest religion in Indonesia, dating to the second century. Hindus in Indonesia are primarily located on Bali, where they represent over 90 percent of the island's population (but only 2 percent of the national population). Hinduism came from India but has been greatly modified by Indonesians to incorporate animist, local, and ancestral practices. Hindus do not believe in one God. Rather, they believe in a number of deities, including many in Indonesia that are different from those found in India. India's caste system has been only loosely adopted by Indonesia's Hindus, and reincarnation is not accepted as it is in the Indian sect of the religion. Hinduism depends less on scripture than Islam or Christian religions, and in Indonesia it is tied more to life cycles (birth, death, puberty, and marriage), art, and rituals.

INDONESIA'S ETHNIC GROUPS

Indonesia is one of the world's most ethnically diverse countries. The archipelago is a virtual United Nations itself, because of the wide variety of cultures scattered across the country. According to Indonesia's 2000 census (the first to include ethnicity since 1930), more than 500 different ethnic groups live in the country. The Javanese are the largest group, followed by Sundanese, Malays, and Madurese. The country also has significant numbers of Chinese, Minangkabau, Batak, Buginese, Balinese, Ambonese, Dayaks, and Sasaks. Many of the ethnic groups, like the Javanese, Balinese, and Ambonese, are named after the island on which their culture developed. This is a reflection of the historical isolation that allowed cultures to develop independently when transportation and communications technology was more limited.

Diversity can be of great benefit to a country. The varied talents, cultures, and backgrounds are drawn together to make a richer society. If ethnic or religious groups stereotype each other, however, division can result, causing mistrust and intolerance. These attitudes can lead to sharp divisions in communities and, in the extreme, they can contribute to ethnic strife and violence if not properly addressed. Indonesia has experienced ethnic tensions between various groups, in addition to the religious divisions and violence noted earlier. This also accounts for the leadership theme of togetherness promoted by Indonesia's leaders, which, for example, in Sukarno's Pancasila, stresses national unity. The fear of ethnic divisions is also a major reason why ethnicity was not tabulated in Indonesia's censuses from 1930 until 2000. Some officials believed that the data would potentially classify people in a way that could be divisive.

The following chart shows major ethnic groups in Indonesia, the percentage of population held by each, and the primary area in which the group is located, and its language.

The Javanese are, by far, Indonesia's largest ethnic group, with over 40 percent of the population and more than 80 million people. Living on their home island of Java and other islands, including Sumatra, Kalimantan, and Sulawesi, the Javanese have a rich cultural heritage that easily adapts to new cultural influences and incorporates these ideas with traditional practices. Javanese tend to cling to their ancestral ties, including the villages from which their families come. Most are also Muslim.

Javanese are well known for their famous *wayang*, or shadow theater. This art form tells Hindu stories in the traditional Javanese language. It focuses upon tales such as Ramayana and Mahabharata, in which good triumphs over evil, usually by drawing upon some mystical forces. These performances are accompanied by music and are performed by actors or puppets. Traditional performances could last all night, but contemporary wayang often last only a couple of hours, to accommodate tourists.

Batik is a beautiful art form that has been created by the Javanese. Dating back to the thirteenth century, batik is an art form in which a hot liquid wax is applied to cotton cloth to create designs. After the wax cools, the cloth is submerged into the dye; this colors all areas not covered with wax. The wax is then removed and the process is repeated with more wax and dye until the artist is finished. Historically, batik was all done by hand; however, modern batik is often done by machine to make the process cheaper for visitors and locals who enjoy the art form.

The Sundanese live primarily in West Java and have some connections to the Javanese. Some linguistic connections allow Javanese and Sundanese to partially understand each other. Family structures are similar, as are economic and

Indonesia's Major Ethnic Groups

Ethnic Group*	Percentage of Population*	Main Location(s)	Language
Javanese	41.70	Java, South Sumatra, Kalimantan, Sulawesi	Javanese
Sundanese	15.40	West Java	Sundanese
Malay	3.45	Coastal areas on Irian, Jaya, Sumatra, Kalimantan, Riau Province	Bahasa Indonesia (form of Malay)
Madurese	3.37	Madura	Madurese
Batak	3.00	North Sumatra	Batak
Minangkabau	2.70	West Sumatra	Minangkabau
Betawi	2.50	Java (Jakarta)	Betawi
Buginese (Bugis)	2.50	Sulawesi	Bugisnese (Buginese)
Bantenese	2.00	West Java, Banten	Banten
Banjarese	1.74	South Kalimantan	Banjarese
Balinese	1.50	Bali, Lombok, West Sumbawa	Balinese
Sasak	1.30	Nusa Tenggara, Lombok	Sasak
Makassarese	1.00	Nusa Tenggara	Makassar
Cirebon	.94	West and Central Java	Cirebon
Chinese	.94	Java, Riau Province	Chinese
Other	15.96	varies	varies

*Indonesia's 2000 Census

political patterns. The Sundanese, like the Javanese, are mostly Muslim. There are also great differences between these cultures, however, that historically have shared the same island. In some ways a rivalry has developed between the two largest ethnic groups in Indonesia, the Sundanese and the Javanese. The Sundanese are more strident with their religious beliefs and each day the call to prayer cries out loudly five times a day from loudspeakers in villages. The Sunda region of Java even attempted to secede from Indonesia in the Darul Muslim separatist rebellion that lasted for 14 years in the mid-twentieth century.

Culturally, the Sundanese and Javanese are similar in some ways. Both cultures, for example, perform the Mahabharata with wooden puppets called wayang goleks. Wonderful traditional dances, such as the sensuous ketuk tilu and japong, classical Sundanese dances, abound in Sunda society. Sundanese musical instruments include the kendang, a local drum that provides a faster rhythm than is usually found in Javanese culture, and the angklung, a unique Sundanese bamboo instrument. The angklung consists of bamboo tubes of varied lengths, which provide different notes and are used much like hand bells.

With the array of cultures in Indonesia, it is impossible to provide a detailed glimpse of each one. Indonesia today is a rich mixture of domestic cultures and imported ones like Malay and Chinese. Traditional cultural practices can still be found alongside those reflecting the blend of ingredients in Indonesia's cultural salad.

Speaking of food, Indonesia also has a rich and varied menu, designed to be practical but also to please the palate. With a historical reference as the Spice Islands, it is not a surprise that Indonesia's foods provide a wonderful assortment of flavors. Spices can include hot chilies, garlic, ginger, and coriander, which are combined with sauces made from coconut milk, peanut sauce, or other base ingredients. The tasty combinations are wonderful and some connections can be made between them and foods of other cultures in Southeast Asia, including Malaya and Thailand.

Sambal Goreng Ayam, or chicken with red pepper sauce, is a Javanese recipe that is representative of spicy red peppers combined with coconut milk. Here is the recipe:

Sambal Goreng Ayam

Ingredients

5 sliced shallots

3 minced garlic cloves

1/2 cup coconut milk

1 teaspoon tamarind soaked in a tablespoon of hot water

1 potato cut into cubes and deep fried

4 cups water

salt and oil as needed for flavor and frying in wok

1 tablespoon soy sauce

6 hot red peppers

1/2 tablespoon sugar

2 chicken breasts

Directions:

Simmer chicken breasts in water until cooked thoroughly. Season to taste with salt while cooking. Let chicken breasts cool, then shred. Save 1/2 cup of broth for use in mixture. Add 1–2 tablespoon of oil to a wok. Stir-fry the shallots and garlic for about four minutes and add the red peppers. Fry the peppers for two minutes and add the chicken. Mix in the coconut milk, soy sauce, sugar, salt, potatoes, tamarind, and 1/2 cup of chicken broth. Simmer for 5–6 minutes and serve with rice.

From isolated cultures locked away on thousands of islands, to its religious diversity dominated by Islam, Indonesia is a country with great human complexity. Blending these human elements together to become one of the great countries in the twenty-first century will be a major political challenge for the Indonesian government. Successfully harnessing the talents of this diverse population will undoubtedly be an important key to Indonesia's future.

CHAPTER

6

Government of Indonesia

Governing Indonesia presents many unique challenges. The vast stretch of islands that comprise this country allows for the development of an extreme range of differing political, social, economic, regional, and religious interests. Because of the isolation, this development can take place without the prying eye of government closely watching the actions and activities of citizens in remote sections of the country. Many isolated places are without adequate law enforcement. The island country's political geography brings to mind the thoughts of Thomas Hobbs, who said, without government "No arts; no letters; no society; and which is worst of all, continual fear and danger of violent death; and the life of man, solitary, poor, nasty, brutish, and short." John Locke said it even more succinctly in noting that "wherever law ends, tyranny begins." These quotations have special meaning for the governance of Indonesia.

THE INDEPENDENT REPUBLIC OF INDONESIA

Indonesia is a republic. This means that the country has a system of government in which voters hold and exercise their power through elected representatives. It is these elected representatives who are responsible for the common good of society.

The birth of the nation of Indonesia was rather traumatic. It came about when Japan announced in September, 1944, that Indonesia would be given independence. The Dutch were furious about this; they wanted to reestablish control over their former colony. Japan had occupied Indonesia for much of World War II, however, and finally surrendered on August 15, 1945. Two days later, on August 17, Indonesia proclaimed independence. With this came the Constitution of 1945, which had been quickly drafted after the Japanese announcement that Indonesia would become independent.

Sukarno was a powerful figure at the end of the war. He advocated a strong presidency and five guiding principles for the new constitution called Pancasila. The five elements included in Pancasila, again, were belief in God, humanitarianism, national unity, democracy, and social justice. These ideas found their way into the Preamble of the 1945 Constitution, which states in the second paragraph,

> . . . the independence of Indonesia shall be formulated into a constitution of the Republic of Indonesia which shall be built into a sovereign state based on a belief in the One and Only God, just and civilized humanity, the unity of Indonesia, and democratic life led by wisdom of thoughts in deliberation amongst representatives of the people, and achieving social justice for all the people of Indonesia.

After independence, Sukarno became the first President of Indonesia. Under his leadership the principles of Pancasila were put into practice in the new country's Constitution. The principle of unity targeted the difficulty of governing a country that stretches over thousands of islands in the world's largest archipelago.

The intrusion of the military into the affairs of government, as described in chapter 4, also presents a challenge to Indonesia. Political leaders must work to keep the military under government rule to avoid having the government held hostage by the country's military leaders. The primary problem with military rule is that government as it is ruled with military might, rather than by voters who elect representatives who are responsible to the citizenry. Sometimes in Indonesia, military rulers have altered the Constitution of 1945 to gain political advantage for selected groups and individuals. Even with these changes, and despite many challenges and numerous recent amendments, the Constitution of 1945 remains the prevalent document in ruling Indonesia.

BRANCHES OF GOVERNMENT

Indonesia's Constitution establishes three branches of government. These include the Executive Branch, the Judicial Branch, and the Legislative Branch. Each of these branches has its own institutions and powers that are established and enumerated by the Constitution. The Constitution also institutes a system of checks and balances, to keep any one branch from becoming too powerful. The Constitution does, however, establish a strong Presidency. This differs from most other democracies, in which the executive branch often has more limited powers.

The Executive Branch

According to chapter 3 of the Indonesian Constitution, the president heads the executive branch. The Constitution also creates a very strong presidency in terms of powers held by the individual holding that office. The vice president is second in command and takes the office of the president, should that position become vacant. The qualifications for president are very basic; they only require that the person be a native-born citizen and that the individual be mentally and physically capable. The office holder also must never have committed an act of treason. Unlike in most countries, there is no minimum age requirement.

Today, a majority of all voting citizens in Indonesia elects the president and vice president directly, by popular vote. As in U.S. elections, the two candidates run on the same ticket. The term of office is five years, and both the president and vice president are eligible for reelection. In earlier years the two executive positions were filled by a vote of the Majelis Permusyawaratan Rakyat (MPR), which meets at least once every five years. This body had unlimited power at one time, but constitutional revisions now require that the President is directly elected by the people.

The office of the presidency carries with it many powers. These include the ability to introduce legislation to the single legislative house, called the Dewan Perwakilan Rakyat (DPR). The president may also develop regulations designed to carry out the laws passed by the legislative branch. He or she is the supreme commander of the Army, Navy, and Air Force and, in agreement with the DPR, can declare war or make peace and treaties. The president can also grant mercy, pardons, and amnesty and restore the rights of individuals. He or she can also grant titles, honors, and distinctions of honor. The office can declare a national state of emergency; this provision provides for additional broad powers that the president may exercise.

The president also appoints the ministers who head various departments of the executive branch. These ministers advise the president and carry out the day-to-day operations of various governmental agencies. The president also has the power to dismiss these ministers as necessary. Indonesia has many national departments, including the Ministries of Agriculture, Transportation, Home Affairs, Defense, Justice and Human Rights, Energy and Mineral Resources, Industry and Trade, Public Works, Finance, Foreign Affairs,

Forestry, Population/National Family Planning, Research and Technology, National Education, Health, Religious Affairs, and even a Ministry for Maritime and Fisheries Affairs. These ministries are tailored to the needs of Indonesia and each falls into one, of three, coordinating ministries. These coordinating ministries are designed to enhance communication, policy planning, and implementation within the following three categories:

- Coordinating Minister for Political and Security Affairs
- Coordinating Minister for Economy
- Coordinating Minister for People's Welfare

Since ministers serve at the pleasure of the president, they usually carry forward the ideas and policies of the person holding that office. Ministers can be replaced at any time deemed appropriate by the president.

The Legislative Branch

Indonesia is somewhat unique among nations in that it has a unicameral, or one house, legislature. This body is called the Dewan Perwakilan Rakyat (DPR), translated to mean the People's Representative Council. This body is established in chapter 7 of the Constitution for the primary purposes of making laws for the country.

Indonesian citizens elect members of the DPR in general elections every five years, at the same time the president and vice president are elected. Elections are constitutionally mandated to be direct, free, secret, and fair. The DPR is required to meet at least once a year. Members may freely express their opinions and submit bills for consideration by the whole body. The DPR may then pass bills that must be jointly approved with the president before they become law. If joint approval is not reached, the DPR may not reintroduce the bill until the following year. This provision gives the president a very strong role in legislation. Bills can become laws without the president's signature if the bill goes unattended and is not signed for 30 days after passage.

Important checks that the DPR has on the executive branch are also included in the Constitution. For example, the DPR is responsible for the budget and has oversight responsibilities that allow it to conduct investigations into matters that they deem important.

Earlier it was noted that the president has broad emergency powers. Under this Constitutional provision, the president may put emergency regulations into place. However, the DPR must approve these regulations at its next session, or the regulations become void and unenforceable. This power helps to restrain the powers of a president who tries to use the emergency powers clause to secure political advantage.

Dewan Perwakilan Daerah

A unique institution that has some relationship to the national government in Indonesia is the Dewan Perwakilan Daerah (DPD). This body can be translated to mean the Council of Representatives of the Regions. This body is also created by the Constitution, but is not a legislative house, as is the DPR. Its role is different and more consultative in nature.

Members are elected to the DPD from every province during the national general election held every five years. The body is required to meet every year, and the membership cannot exceed one-third of the membership of the DPR. The primary power of the DPD is quite limited, but it serves a useful purpose in providing a forum for discussions about the relationships between the central government and local governments. The DPD can propose bills to the DPR that are related to regional autonomy, the relationship between the local and national governments, the political expansion or merger of regions, and the management of natural resources.

Their primary activities are proposing, discussing, and overseeing the implementation of the bills that are proposed to the DPR.

Majelis Permusyawaratan Rakyat

Another unique body established in chapter 2 of Indonesia's Constitution is the Majelis Permusyawaratan Rakyat (MPR). This is the People's Consultative Assembly, which is required to meet at least once every five years in the capital city of Jakarta.

The MPR has very important Constitutional duties. These include the ability to amend the Constitution and, if necessary, to remove the president or vice president from office. A majority vote is required to act, and the ease of this requirement has contributed to the frequent amending of the Constitution in recent years. A ceremonial responsibility of the MPR is that it formally inaugurates the president and vice president.

The Judicial System

A court system is established in chapter 8A of the Constitution. The Supreme Court is the country's highest judicial body. It is charged with the task of hearing cases of law and providing justice. These cases may relate to regulations and laws as they are applied and interpreted in Indonesia. Other lower courts hear cases for the first time and also are used for specialized matters such as religious affairs or military matters. Surprisingly, the Supreme Court does not have primary or first-time jurisdiction over constitutional questions that arise. That responsibility is given to another court, called the Constitutional Court.

Candidates for the Supreme Court are nominated by a special Judicial Commission that reports and recommends to the DPR. Members of the Judicial Commission are appointed by the president, but the legislature then approves or disapproves the candidates. Successful candidates are then formally

Like the United States and other democracies, Indonesia has a constitution and a three-part government, with executive, legislative, and judicial branches. Here, Muslim cleric Abu Bakar Bashir, on trial for treason, sits in a Jakarta court in 2003.

appointed to the Court by the president. The highest position on the Supreme Court is the chair; this person is elected by a majority vote of the justices.

There is also a special Constitutional Court that hears cases that might be in conflict with the Constitution or need Constitutional interpretation. This court has nine justices, three each of whom are nominated by the president, DPR, and Supreme Court. The nominations process serves as another balance of power between the three branches of government. The president formally appoints the justices to the Constitutional Court, and the justices elect their own chair and vice chair. This court also has the power to dissolve political parties if they deem the action appropriate. Some other countries, like Germany, which has banned the Nazi Party, also have provisions similar to Indonesia's. Obviously, a law like this could be abused in the hands of a rogue court and a dictatorial leader.

ROLES OF CITIZENS

The role of the citizen is vital in Indonesia as it is in other democracies. A number of protections and duties are outlined in chapters 10 and 11 of the Constitution, which details these issues in the section on citizens and residents. Like most democracies, Indonesia stresses the equality of all citizens under the law. Citizens have the right to work as well as the right and duty to participate in the defending of the country.

A number of vital rights and freedoms are protected and listed in the Constitution to preserve human rights and liberty. Among these are:

- The freedom to associate with others and assemble
- The freedom to express written and oral opinions
- The right to live and defend his or her existence
- The right to establish a family and have children
- The right to live and be protected from violence or discrimination on any basis
- The right to an education, including science, technology, arts, and culture
- The freedom to choose and practice religion of his or her choice
- Freedom from torture and inhumane treatment
- The right to own personal property

These freedoms may seem very positive and strong. In practice, however, there have been many problems because the actual laws designed to implement these Constitutional protections have often weakened the written intent. For example, in the Constitution, "the freedom to assemble and express written and oral opinions" has been weakened. A provision was in the Constitution adds the statement, "shall be regulated by law."

Examples of citizen's rights being diluted or abused can be shown time and again. For instance, during the early years (including the Suharto regime), the plight of political prisoners was desperate. Many political dissidents were persecuted by the government and military. More recently, after the end of the Suharto era, great strides have been made in promoting freedom of the media. This has created a great expansion of media outlets for both print and electronic journalism. In 2003, however, Amnesty International expressed concerns over the growing tendency of the government to bring cases against journalists. There is a fear that new restrictions will be placed on the media. Additionally, Amnesty International reported that hundreds of people in the Papua region, also referred to as Aceh and Irian Jaya, disappeared or were allegedly tortured by government security forces. Thus, while progress has been made in having Constitutional rights put into practice, more work remains for Indonesia.

Citizens also have limited duties assigned by the Constitution, such as the requirement to defend the country. Additional duties include respecting the human rights of other citizens and respecting the rule of law and the government.

LOCAL GOVERNMENTS

Power and authority in Indonesia are separated by the three branches of government at the local as well as national level. In addition to the national government, separate governments function at the province, regency, and municipality

Although Indonesia is a democracy, the country has had a history of abusing or ignoring the rights of its citizens. The media, for example, is often restricted, and here, journalists rally to support fellow reporters who have been held hostage or killed on the job.

levels. Each of these levels of government has its own areas of jurisdiction and responsibility.

Indonesia's provinces are similar to states in the United States. Their executive branch is headed by a governor. *Kabupaten* is the Indonesian term for "regency," political units that are somewhat similar to counties in the United States. They are headed by a *bupati* or regent. Even smaller is the *kota,* or municipalities, which are headed by a *walikota* or mayor. Each province, regency, and municipality has a legislative body that is usually called a Regional People's House of Representatives. These bodies are similar to the national legislature in that they make laws and public policies for the jurisdiction for which they are responsible. Members of these bodies and the local executives are elected in general elections every five years.

The complexity of governing Indonesia continues to evolve as political thinking develops further and the Constitution undergoes regular review and amending. As the country struggles with separatist movements, terrorist attacks, and other divisions, many see the need for leaders with a strong hand. Without strong checks and balances, the country can slip back into old patterns of strong leaders and military intervention. These historical patterns are still close and threatening. Only a vigilant citizenry and watchful media can keep the country on its present path toward becoming a true democracy, with the rights of all citizens protected in the spirit of unity described in Pancasila.

CHAPTER

7

Indonesia's Economy

Indonesia's economy is as complex as other aspects of the country's physical, historical, and cultural geography. The fact that the country is blessed with many important resources is evident in the country's history. For centuries, Indonesia served as a rich treasure trove for colonial powers, including Portugal, the Netherlands, and Japan. Despite its rich natural endowment, the country's economy has long failed to meet its potential. Indonesia's economic development has been undermined by a long series of disruptive secessionist movements, sharp ethnic divisions, rampant corruption, and mindless terrorism. Added to these woes, of course, is the country's physical geography—thousands of islands lying scattered over thousands of miles. The country's banking system and its relationship with the International Monetary Fund impose other problems. These instabilities and others have a cumulative effect of discouraging foreign investors.

Indonesia has a huge labor force. Nearly 100 million people are available for or engaged in work. What types of jobs do they hold? A lot of the economy is still very traditional, with 45 percent of the population engaged in agriculture. Service industries employ 39 percent of the work force, and only 16 percent of all working Indonesians are employed by industry. More than 10 percent of the working-age population is unemployed. Unlike some of Indonesia's neighbors, such as Singapore and Malaysia, manufacturing has not advanced as rapidly. This is caused in part by Indonesia's uncertain political and economic conditions.

AGRICULTURE

Agriculture provides both products and jobs in Indonesia. With the rugged terrain, only about 10 percent of the land is now farmed, though another 20 percent has agricultural potential. Most commercial crops are grown on large estates, but people often farm on small plots or have their own home gardens. Important plantation crops include rubber, cocoa, coffee, palm oil, coconuts, and tobacco. Key food crops are rice, peanuts, cassava, soybeans, sweet potatoes, corn, and vegetables. Timber is also grown as a "crop." Spices such as cloves and nutmeg continue to be important crops, as well. Animals raised for production include poultry, pigs, and cattle. Fishing is a very important coastal endeavor.

Rice, the staple of the Indonesian diet, is the most important crop. It is grown in a paddy, a flooded field in which farmers wearing cone-shaped hats wade knee-deep in water to tend their crop. Green, terraced hillsides of flooded rice paddies present a visual of rice during the growing season. Several decades ago, Indonesia was a rice importer. Since the arrival during the 1980s of "green revolution" agricultural technology and methods, the country has been self-sufficient in rice production.

Slash-and-burn techniques of shifting subsistence cultivation are still used in much of the country. Trees and other vegetation are slashed, allowed to dry, and then burned in order to open new land to farming. This results in unwanted air pollution, and often the

Rice is the most important crop in Indonesia's agriculture industry, mainly because it is a regular part of Indonesians diets. Here, a farm worker gathers rice in a paddy, or a flooded field.

land that is burned is not extremely suitable for agriculture over the long term. In 1997, agricultural burning in Indonesia created a vast cloud of atmospheric pollution that extended over much of Southeastern Asia.

MANUFACTURING

Manufacturing involves only a small percentage of Indonesia's workers. The country has not moved as quickly forward industrially as have some other Asian nations. This is attributable in part to the country's perceived political instability—secessionist activities and various ethnic and religious divisions make potential investors wary. Today, however, employment in manufacturing is increasing, whereas jobs in the agricultural sector are declining in number. Even with only 16 percent of the population engaged in manufacturing, this sector of the economy is very productive. In fact, it is responsible for 42 percent of the Gross Domestic Product (GDP). Oil and natural gas production are the most important industries. Agriculture only accounts for 17 percent of the GDP, and services account for the other 42 percent.

Major industries after petroleum and natural gas include the manufacturing of textiles, shoes, and clothing. Cement, lumber products, steel, and other metal products also are important. So, too, is the manufacture of fertilizer, cigarettes, appliances, and chemicals. Over 80 percent of the manufacturing takes place on Java and Sumatra, where large pools of labor are available. Efforts are being made to develop more small industries, particularly those that could become exporters. If successful, this will diversify the economy further and fuel industrial growth.

Indonesia has been blessed with natural resources that have long served as a major underpinning of the country's economy. Oil and gas are vitally important to the country's economic health. Two-thirds of the petroleum is exported, and Indonesia is one of the world's largest exporters of natural gas, much of which goes to Japan. As a major oil producer, Indonesia is a member of the Organization of Petroleum Exporting Countries (OPEC). (For additional information on this organization, enter "opec" into any search engine and browse the sites listed.) Other resources include the metals nickel, bauxite, iron, and manganese; building materials such as marble, limestone, clay, gypsum, asphalt, and asbestos; and kaolin, phosphate, iodine, and sulfur. Indonesia is a major exporter of these products and resources. Indonesia exports mainly to Japan, but also to the United States, Singapore, South Korea, China, Taiwan, and

The manufacturing sector accounts for almost half of Indonesia's Gross Domestic Product, and natural gas production is one of its most important industries. Here, fishermen float past natural gas storage tanks in the Aceh province.

Australia. Imports come primarily from Japan, South Korea, and Singapore. They are followed in order by China, the United States, and Australia. Primary imports include food products, chemicals, fuel, and machinery and equipment.

Tourism is another important industry for Indonesia—the islands present a virtual tropical paradise for the adventurous traveler. Beautiful mountains skirted by clouds, sandy beaches, and fascinating cultures provide visitors with an incredible array of features to explore. With its favorable climate, people from across the world have traveled to Bali and other islands. They come for rest and relaxation, and also for trekking adventures, scuba diving, surfing, and a wide variety of other activities. Many of Indonesia's service jobs are found in the tourist industry.

INTERNATIONAL ALLIANCES

Indonesia is a founding member of the Association of Southeast Asian Nations (ASEAN). This intergovernmental organization was started in 1967 with four other nations, including Malaysia, Singapore, Thailand, and the Philippines. ASEAN was formed to foster cooperation in economic, social, cultural, technical, educational, and other fields of mutual interest and concern. The organization also strives to promote regional peace and stability through respect for justice and the rule of law, and adherence to the principles of the United Nations Charter. ASEAN's economic and social development efforts have been of great benefit to Indonesia. Today, there are 10 members of ASEAN: Vietnam, Cambodia, Brunei-Darussalam, Laos, and Myanmar have also joined. With exception of the brief economic crisis experienced during 1998–1999, Indonesia has steadily increased its GDP by a healthy 3 1/2 percent, to nearly 8 percent per year. ASEAN membership has helped Indonesia shift its economy from a simple agrarian foundation to becoming an important player on the world economic stage.

COMMUNICATION AND TRANSPORTATION

It is no surprise that communication and transportation between the people of Indonesia and their neighbors has been challenging. The ocean and rugged island terrain have proven to be difficult and costly barriers to overcome. During the past century, a number of major technological advances have made access much easier. A limiting factor has been and continues to be the costs associated with developing communication and transportation linkages.

Connecting the islands is a difficult, if not impossible, task. A variety of ships, ferries, and other seagoing vessels are used between many islands, but by no means between all of them. Air travel is provided by the national airline, Garuda Indonesia Airways, and smaller domestic airlines, such as Merpati Nusantara Airlines, Pelita Air Service, Bouraq Indonesia Airlines, and Aerowisata International (a subsidiary of Garuda). Other airlines, such as Air Paradise International, cater to tourist traffic, which, in this case, is between Bali and Australia. Unfortunately, the cost of air travel is high for the average Indonesian. Both of these factors limit inter-island travel. Many Indonesians, in fact, never leave their island of birth.

On the islands, travel is usually by road or, on the larger islands, railway. As one would expect, the rugged terrain hampers travel and can make road and rail construction difficult and costly.

In the late 1990s, only 5 1/2 million people had landline telephones. At that time, the cost was prohibitive to most Indonesians. Today, however, cell phones are appearing in greater numbers. Cell phones employ satellite technology, rather than using expensive landbased lines and switching systems. The rapid increase in the use of cell phones makes it possible for Indonesians to communicate more conveniently with one another and with the outside world. Indonesia also has a fairly efficient postal service.

Traditional media systems, especially radio, have served as the primary news provider for Indonesians on a daily basis. At the dawn of the twenty-first century, more than 31 million radios and about 14 million televisions were in use. The country has more than 700 radio stations and about 40 television stations, all privately owned.

The Internet is slowly gaining in popularity. In 2000, the country was estimated to have only 400,000 users and 24 providers. Connecting to the outside world by Internet is still

Building modern means of transportation and communication has been difficult in Indonesia because of the number of islands, the distance between them, and the rugged landscape. However, during the last half century technology has improved, and the number of cell phones, radios, and televisions in use has steadily increased. Here, Indonesians use public telephones at a shopping mall in Jakarta.

limited, although cyber cafes are becoming more common and local Internet businesses are springing up on the islands.

INDONESIA'S CURRENCY

Indonesia's currency is the Rupiah. In 2004, a person in Indonesia could be a millionaire with only about $125.00 in

U.S. dollars! Credit cards are not frequently used in this country, and checks are not accepted—fraud is a problem. Thus, Indonesians and visitors carry and use the Rupiah. At the beginning of 2004, one U.S. dollar converted to about 8,500 Rupiah. The largest currency denomination was a 100,000 Rupiah bill worth about $12 in U.S. dollars.

Because the government sometimes deflates the Rupiah against the dollar, Indonesia's currency also has contributed to the instability of the economy and daily life of its people. Major devaluations took place in 1973, 1983, and 1986. In 1983, virtually overnight, the government changed the official exchange rate from 620 to 1,000 Rupiah to the dollar. Deflation has a devastating effect on Indonesians who hold the currency when it is devalued because prices of products and services rise immediately and drastically to reflect the Rupiah's lowered value. With devaluation, the price of staples such as rice may double over night. At the same time, local people who are paid in foreign currencies experience a windfall after devaluation.

Since 1986, the government has floated the Rupiah against the dollar and other international currencies. This has worked well with the exception of 1997–1998, when a major economic crisis hit the country. During the crisis, Indonesians and their currency were taken on a roller-coaster ride, with value soaring from 2,450 to 17,500 to the dollar and then dropping back down to 7,750 by the end of 1998.

At the beginning of the twenty-first century, Indonesia is moving forward at a hopeful pace. The country has benefited from its membership in ASEAN and its own more stable political situation. It has entered into new and supportive agreements with the International Monetary Fund (IMF) and others economic groups. New technologies in communications and transportation have linked Indonesia closer to the outside world. These factors, in combination, can help Indonesia become one of the world's leading economic players. With the world's fourth-largest population and a strong natural resource base, the country is well poised to improve life for its citizens, although political insurgents, corruption, terrorism, and ethnic and religious strife exist in the wings to threaten economic progress. As separatists, terrorists, and other divisions within the country threaten the stability of the government, they also inspire fear in tourists and potential foreign investors. News of the Bali nightclub bombing and the 2003 bombing of a Jakarta hotel were carried around the world. Such reports have had a negative effect on the important tourism industry. Whether the effect is of short- or long-term duration has yet to be determined. What we do know, however, is that Indonesia's future economic vitality remains dependent upon increased political stability and further reduction in corruption.

CHAPTER

8

The Islands

It is nearly impossible to conceive of more than 17,000 islands strewn in a wide swath across the area connecting the Indian and Pacific Oceans, yet these are the islands that comprise the country of Indonesia. All the islands have a story to tell, but we can't visit each of them in one small book. Instead, we will take a closer look at some of the more significant islands and examine the human and physical characteristics that they exhibit.

JAVA

Java was formed by 38 volcanoes, 13 of which are still considered active. Fortunately for the island's residents, no major eruptions have occurred since the mid-nineteenth century. This humid tropical island is Indonesia's historical, economic, political, and cultural heart. It also occupies a central location in the archipelago. Although 115 million people, or nearly one-half of

Indonesia's total population, call this island home, the island only contains about 1 percent of the country's land area. This means that the population density is very high, with about 1,400 people per square mile (870 per square kilometer). If Java were an independent country, this would be the second-highest density in the world, behind only Bangladesh. In human terms, this translates into noisy, crowded urban areas with congested traffic, extensive pollution, and the many other problems that compressed cities face.

Jakarta, Indonesia's capital, is located on west Java. In its early history, the city was called Jayakarta. Destroyed by the Dutch East Indies Company under Jan Pieterszoon Coen in 1619, the city has reemerged to become one of the world's largest urban centers. At the end of World War II, the city had fewer than one million people. By the beginning of the twenty-first century, the population had surged to over 10 million.

Jakarta is truly an amalgamation of Indonesia as a whole. The city has a mixed population that includes the richest of the rich and the poorest of the poor. Skyscraper office buildings decorate the city's skyline, whereas slums hide in their shadows. Jakarta presents a stark contrast between wealth and poverty.

Jakarta enjoys some beautiful architecture: In addition to the capitol building, the city boasts the Istiqlal Mosque, the largest in Southeast Asia, as well as national museums and the national headquarters of many businesses. Jakarta is where Indonesian popular culture begins and is broadcast across the country by the media headquartered in the city. The city is the country's airline hub and has the major seaport at Sunda Kelapa Harbor. The port has been active since the sixteenth century, when Europeans docked their ships in search of spices. Today, the harbor serves as a major container port for materials from other islands and for trade with the world.

The government has divided Java into four provinces. These include West Java, Central Java, East Java, and Banten, which became a province in 2000. Jakarta is designated as the capital city district,

Jakarta is the capital of Indonesia and one of the world's largest cities. It is the center of Indonesia's popular culture, government, and economy, and features many architecturally impressive buildings, like the skyscrapers seen in this overview of the city.

and another special region, a sultanate called Yogyakarta, serves as the cultural center for the entire island and an important academic center. Each of the provinces has its own cultural heritage and personality. Javanese is spoken in central and eastern Java and Basa Sunda is spoken in western Java by about 25 percent of the island's population. The roots of this language are found in the Sunda Kingdom, which dates back to the fifth century.

BALI

Located to the east of Java is the tropical island tourist destination of Bali, with its world-famous beaches. Long a vacation haven for Australians, Americans, and Europeans, Bali has a culture different from most of Indonesia. Here, the roots of India are deeply set, and Hinduism is more prevalent than on other islands. Ninety-five percent of the population is Balinese Hindu. Bali has also taken and modified India's caste system into a less strict division of classes. As with other aspects of Hinduism, Bali has localized the religion into its own version of the faith.

Bali, with an area of 2,192 square miles (5,620 square kilometers), is a little larger than the state of Rhode Island. Nearly 3 million residents are crowded into this picturesque setting. Denpasar, with a population of about 400,000, is Bali's capital and largest city. Denpasar is a rapidly growing city, with increasing traffic noise, pollution, and other urban problems. It also retains popular attractions such as tropical gardens, museums, and art galleries. Nearly a million visitors a year are attracted to the island, and most spend time in Denpasar. About 10 miles north of Denpasar is the city of Ubud. This relaxed city serves as Bali's cultural center, with additional museums, art galleries, and cultural performances. Graceful dancers perform traditional Balinese dances for visitors in Ubud's cultural centers, and the local art is Bali's finest.

Bali's history can be traced back to written records from the ninth century, but it is certain that people occupied the island long before that time. Hindu influences started appearing around the eleventh century. Because the island held little of value, European powers all but ignored it until the Dutch arrived in 1597. These early Dutch crew members, however, recognized the great natural beauty that Bali offered.

Bali was under Dutch control until the Japanese arrived during World War II. When Indonesia became independent after the war, the island became a part of the new country. With its Hindu traditions, there has often been strong resistance to the influences of Islam on the island. This made the island a convenient target for Jemaah Islamiyah terrorists, who bombed the Sari Club near Kuta Beach in October, 2002, and killed over 200 people, mostly Westerners and Australians. The Jemaah Islamiyah organization is affiliated with al Qaeda, and the bombing has negatively affected tourism, Bali's most important industry.

SUMATRA

Sumatra is the second-largest island in Indonesia and home to nearly 37 million people. The word *sumatra* is a variation of the Latin phrase *suma terra,* which means black earth. Sumatra has a land area of 183,000 square miles (473,970 square kilometers), slightly larger than California. It is the westernmost island in the archipelago. Many smaller islands dot the waters surrounding Sumatra. The strategic Strait of Malacca separates the island from the Malay Peninsula, which is close by, to the northeast. Running down the spine of the island is the Bukit Barisan range, which has peaks reaching over 11,000 feet (3,350 meters). The range is volcanic in origin, and 20 of its 50 identified volcanoes are still active.

Sumatra is Indonesia's resource treasure house. Nearly 70 percent of the country's economy is generated on Sumatra, primarily because of its oil fields, gold, silver, bauxite, tin, and lignite coal production. With its fertile black soil, the island is also Indonesia's leading area of agricultural production. Most of the country's rubber is gathered here, but the island also

Sumatra is the second-largest island in Indonesia, but is less urbanized than many of the other islands. It holds most of the country's resources, is the leading area of agricultural production, and has large areas of rain forest that are home to endangered species.

produces substantial amounts of pepper, rice, corn, coffee, tea, timber, palm oil, sisal, sugar cane, and tobacco.

In contrast to Bali and Java, Sumatra is lightly populated. Natural plant life and wildlife is much more common in the less-urbanized atmosphere of the island. Because of this, the island still has vast reserves of rain forest. These areas provide a habitat sanctuary for many endangered species of animals such as tapirs, tigers, leopards, rhinoceroses, and elephants.

Because of its size, Sumatra is administratively divided into eight provinces. Local populations primarily include the Acehnese, Batak, Minangkabau, and coastal Malays. Islam is the major religion. Various small tribes, some numbering only a few hundred people, also strive to preserve their traditional cultures in isolated communities.

Marco Polo visited Sumatra on his way home from China in 1292. The islands had been populated for a very long time, however, with Arabs and Indians arriving long before Polo. Because of Sumatra's riches, the Portuguese, Dutch, British, and Japanese all had interests in the island and each held all or parts of the island at different times before Indonesian independence.

Sumatra today also has some problems, especially on the northwestern tip of the island, which is home to the Acehnese. The area of Aceh has been in frequent rebellion against the government since the 1950s. A full-scale rebellion occurred in 1958, but it was put down by Indonesian forces. Since that time, there have been frequent guerrilla attacks, and many demonstrations have taken place. The Muslim Acehnese seek to install Islamic Sharia law and gain their independence from Indonesia. The country recently gave Aceh more local autonomy, but resistance to the national government has continued in 2004, resulting in martial law being instituted in Aceh.

Although short stories have been told of only three islands in this chapter, the stories provide an in-depth look at some of the diversity found within Indonesia. There are thousands of additional island stories to tell and each contributes to the complexion and composition of Indonesia. (For additional information on other Indonesian islands, enter "islands of Indonesia" into any search engine and browse the sites listed.)

CHAPTER

9

Indonesia's Future

Shakespeare wrote "All the world's a stage, and all the men and women merely players: They have their exits and their entrances." Applying this quotation to Indonesia's past, present, and future provides a lens for examining an extremely complex drama. In Indonesia, even the stage, the landscape, is alive with earthquakes, volcanoes, typhoons, and the sea constantly brushing against the islands' shores. Human players have been on Indonesia's stage for thousands, perhaps even millions, of years. Unique cultures developed on isolated island stages, where local indigenous cultures flourished in presenting their cultural interpretation of daily life.

Foreign players have appeared on the stage in leading roles with the entrances of the Portuguese, Dutch, Spanish, British, and Japanese. Each has impacted Indonesia's past and has affected cultures and the course of history in the region. Because more than 6,000 of the islands have human inhabitants, the thousands of individual stages and dramas, taken together, create a somewhat chaotic setting for a country's foundation.

Indonesia today is a country challenged by its own complexity. Its complexity also presents tempting opportunities, if the country can harness its own potential. Predicting the future is a precarious endeavor—many possible images come into view when looking ahead. Each scenario is dependent upon certain assumptions that may or may not happen. Let's review some of the foundations that exist now in Indonesia.

Indonesia has a rich natural resource base—more than 17,000 islands—and possesses the world's fourth-largest population. The government has wavered between democratic regimes and those exercising more strict presidential control in efforts to hold the complex nation together. When rulers become too authoritarian, protests increase and the government then turns over and usually becomes more democratic, as it did after Sukarno and Suharto. If the central government is not authoritarian enough, pieces of Indonesia break away from the country. This happened recently with East Timor and could result from the current separatist movement in Aceh. This political paradox places continual pressure on the government to find a balance.

Indonesia also is sharply divided by regional, ethnic, and religious differences. The country could fracture, politically, into many smaller countries, as happened to the former Soviet Union and Yugoslavia after the end of the Cold War. The assumption that the government can find the proper balance between central control and democracy may help to facilitate a brighter and united future for Indonesia.

The goals of God, humanitarianism, national unity, democracy, and social justice that Sukarno set forth in Pancasila have been continued under Suharto and present leaders. These five ideas provide a basic plan for the future. The cry for national unity is an obvious response to the potential divisions existing in the country. Social justice yields to citizens the harvest of equality and fairness before the law, regardless of ethnicity, gender, or religion. This can mean that Indonesians have hope and equality of opportunity. Development of a true democracy has been elusive, however; leaders have grabbed more power when civil disobedience or separatist movements have threatened the nation.

In response, new generations of Indonesian students are being taught more about the role of a citizen in a democratic society. Project Citizen, for example, has been active in the country. It is a program developed by the Center for Civic Education in Calabasas, California, and supported by the U.S. Agency for International Development (USAID). This program helps Indonesian high school students identify local problems and create public policies to present to local governments for possible passage. The program also increases tolerance and understanding of others. Through programs like Project Citizen, students practice the skills, apply their values, and use knowledge that will help them to improve their community and potentially their country. Programs like this will assist in strengthening democracy in Indonesia.

Other important challenges, such as rampant corruption, face Indonesia. International Corruption Watch (ICW) lists Indonesia as one of the worst countries in the world for corruption. Transparency International identifies the country as Asia's worst in this category. According to the World Bank, "corruption has been and continues to be a problem in Indonesia." A high percentage of the funds from foreign aid leak out to government staff and elected officials. Under Suharto, corruption seemed to be a problem only at the level of top officials. In recent years, however, the problem has affected officials at much lower levels. Some outsiders believe that corruption is a trait ingrained in Indonesian culture. Whatever the cause(s), the effect has been to cause outside lenders and investors to be wary of having funds

Indonesia is a country rich in culture and resources. However, it is also challenged by political instability and regional differences. If the country can continue on a path of democracy and unity, there is hope for its future and its youth, like these children marching during an Independence Day parade in Rantepao.

stolen or skimmed off if they go to Indonesia. Thus, this issue is not only a problem of human values, but it has become significant in political, social, and economic realms as well.

Environmental challenges also factor highly in Indonesia's future. The country possesses 10 percent of the world's forests and the third-largest rain forest in the world. These are rapidly being deforested at a rate that was the fastest in the world during the late 1990s. Air pollution from cars and urban noise pollution are also major environmental challenges. Water pollution of the sea-lanes around Indonesia is an important issue, together with many other environmental problems. Regulations to address these issues are presently scarce. New laws and their honest enforcement will be needed to improve the environment.

Although population growth has slowed, the sheer number of people has led to major problems, especially in urban centers on Java and some areas on Bali. The high population density, combined with inconsistently applied environmental protection laws, serve to decrease the quality of life for residents. Population growth can further exacerbate these problems.

In the summer of 2003, terrorists struck Indonesia again when the famous Jakarta Marriott Hotel was bombed. This tragic event killed 13 people and injured another 150; Jemaah Islamiyah, a group linked to al Qaeda, claimed responsibility. Members of this terrorist organization trace their roots back to Muslims who opposed Suharto. The group is on a rampage against the United States and other Westerners, with the goal of establishing an Islamic state in Southeast Asia. Relations with the West were not enhanced when in 2003 Indonesia's vice president, Hamzah Haz, called the United States the "king of terrorists" because of the U.S. involvement in Iraq. Haz's comments reinforce the perception held by many Westerners that at least some people in Indonesia's government are too close to and supportive of extremists in the country.

Many challenges exist for the Indonesian people and their government as the country's drama is played out on the world's stage. Watching the developments in coming years will be important. The country is too significant to ignore, because its impact on Southeast Asia and the world is very large. Continued development of democratic and Pancasila ideals would appear to lead the country on a path of hope. Other paths appear fraught with pitfalls and troubled destinies. The citizens of Indonesia and their government will determine the drama that will take place. The world is hoping there will be a good ending.

Facts at a Glance

Country Name	Republic of Indonesia
Location	Southeast Asia, an archipelago between the Indian and Pacific Oceans
Capital	Jakarta
Area	705,192 square miles (1,826,440 square kilometers); about three times the size of Texas
Land Features	Coastal lowlands, with interior volcanic mountains
Climate	Tropical, hot, humid, with more moderate temperatures in the highlands
Coastline	33,999 miles (54,716 kilometers)
Natural Hazards	Occasional floods, severe droughts, tsunamis, earthquakes, volcanoes, forest fires
Land Use	Arable land: 9.9% Permanent crops: 7.2% Other: 82.9%
Natural Resources	Petroleum, tin, natural gas, nickel, timber, bauxite, copper, fertile soils, coal, gold, silver
Environmental Issues	Deforestation; water pollution from industrial waste, sewage; air pollution in urban areas; smoke and haze from forest fires and agricultural burning.
Population	235,000,000 (July 2004)
Population Growth Rate	1.52 %
Total Fertility Rate	2.5 children born per woman
Life Expectancy at Birth	Total population: 68.94 years Female: 71.47 years (2003) Male: 66.54 years
Ethnic Groups	Javanese 41.7%, Sundanese 15.4%, Coastal Malays 3.45%, Madurese 3.37%, other 26%
Religion	Muslim: 88% Protestant: 5% Roman Catholic: 3% Hindu: 2% Buddhist: 1%

Languages	Bahasa Indonesia (official), hundreds of local dialects, English, Dutch
Type of Government	Republic
Executive Branch	President who serves as head and chief of state
Independence	August 17, 1945
Administrative Divisions	27 provinces, 2 special regions, and a special capital city region (Jakarta)
Currency	Rupiah
Industries	Petroleum and natural gas; textiles, apparel, and footwear; mining, cement, chemical fertilizers, plywood; rubber; food; tourism
Unemployment Rate	10.6% (2002)
Primary Exports	Oil and gas, electrical appliances, plywood, textiles, rubber
Export Partners	Japan 19.2%, U.S. 14.5%, Singapore 11.6%, South Korea 6.6%, China 5.6%, Taiwan 3.7% (2002)
Imports	Machinery and equipment; chemicals, fuels, foodstuffs
Import Partners	Japan 18.2%, South Korea 9.6%, Singapore 8.4%, China 7.9%, U.S. 7.6%, Australia 5.0% (2002)
Transportation	Railroad: 4,013 miles (6,458 kilometers) Highways (total): 342,700
Highways (paved)	98,593 miles (158,670 kilometers)
Airports	631
Ports and Harbors	Cilacap, Cirebon, Jakarta, Kupang, Makassar, Palembang, Semarang, Surabaya

History at a Glance

1,700,000 B.C. to 500,000 B.C.	Java man existed.
40,000	Early inhabitants arrive from Malay Peninsula.
2,500	Pottery and stone tools used.
A.D. 100	Hindu kingdoms begin appearing.
300	Arab trading with communities on the archipelago.
425	Buddhism arrives in Sumatra.
600	Beginning of Srivijaya kingdom on Sumatra.
825	Borobudur temple is finished.
1292	Marco Polo briefly stops in Sumatra and Java on his travels home.
1300	Islam strongly entrenched on many of the islands.
1511	The Portuguese arrive in Indonesia.
1560	Spanish arrive at Manado.
1579–1580	The Englishman Sir Francis Drake visits various islands in the archipelago.
1596	Dutch first arrive in four ships.
1602	Dutch East India Company (VOC) chartered.
1611	British set up trading posts on various islands.
1618	Jan Pieterzoon Coen is appointed Governor General of the VOC.
1629	Coen dies during Sultan Agung's effort to retake Jakarta; Coen's VOC forces destroy Batavia (Jakarta) by burning it to the ground.
1646	Sultan Agung dies.
1732	Malaria epidemic breaks out in Jakarta.
1784	Treaty of Paris signed, ending the Dutch war with England; islands are opened to free trade and not just the VOC monopoly.
1799	Dutch East India Company (VOC) goes bankrupt.
1811	Thomas Stamford Raffles is appointed Lieutenant Governor.
1811–1816	The British rule Java and Dependencies (Indonesia).
1815	Eruption of Tambora Volcano.
1816	The British return Indonesia to Dutch control.
1817	Pattimura leads a revolt against the returning Dutch in Ambon.

1825–1830	The Java War rebellion against Dutch is led by Prince Pangeran Diponegor.
1845	Vanilla production begins on Java.
1859	The Dutch end slavery in the Netherlands Indies (Indonesia).
1864	First postage stamp issued by the Netherlands Indies.
1872–1895	Batak wars in northern Sumatra.
1873–1874	Dutch war with Aceh (guerrilla warfare continues for decades).
1883	Eruption of Krakatoa.
1906	Bali falls to the Dutch.
1908	Budi Utomo founded by Soetomo.
1927	Indonesian Nationalist Party (PNI) founded by Sukarno.
1933	Sukarno, Hatta, and other PNI leaders arrested and imprisoned by the Dutch.
1940	The Netherlands fall to Germany.
1941–1945	Japan conquers the Netherlands Indies and occupies the colony.
1942	Sukarno escapes from Dutch imprisonment.
1943	Allies begin the retaking of Indonesia.
1945	Japan surrenders to Allies; Sukarno declares Indonesia's independence from the Dutch; four-year war with Dutch begins; new constitution implemented; Sukarno declared president.
1949	Dutch relinquish control of Indonesia.
1950	Indonesia admitted to the United Nations; Garuda Indonesian Airlines founded.
1956	Hatta resigns as vice president.
1957	Sukarno introduces the idea of "Guided Democracy."
1962	Irian Jaya (West Papua) is joined to Indonesia because of UN action; Indonesia joins OPEC.
1963	Sukarno declared president for life.
1965	Indonesia withdraws from the United Nations.
1966	Sukarno removed from power and Suharto assumes leadership.
1967	Association of Southeast Asian Nations founded by Indonesia and four other countries; Indonesia rejoins the United Nations.

History at a Glance

1971	General Suharto elected president for the first time.
1975	Portuguese leave East Timor; Indonesia invades East Timor.
1976	East Timor becomes a province of Indonesia.
1997–1998	Indonesia experiences a financial crisis.
1998	Suharto resigns from office.
1999	East Timor breaks away from Indonesia.
2001	Impeachment of Abdurrahman "Gus Dur" Wahid; Megawati Soekarnoputri becomes president.
2002	Bali nightclub bombed by terrorists affiliated with al Qaeda and more than 200 are killed. East Timor gains full independence.
2003	Jakarta Mariott Hotel bombed by Jemaah Islamiyah.

Alatas, Syed Farid, and Farid Alatas. *Democracy and Authoritarianism in Indonesia and Malaysia: The Rise of the Post-colonial State.* New York: Palgrave Macmillan, 1997.

Alexander, Paul, ed. *Creating Indonesian Cultures.* Sydney, Australia: Oceania Publications (University of Sydney), 1989.

Anwar, Dewi Fortuna. *Indonesia in ASEAN: Foreign Policy and Regionalism.* New York: Palgrave Macmillan, 1994.

Bourchier, David, and Vedi Hadiz, eds. *Indonesian Politics and Society: A Reader.* New York: Routledge, 2001.

Cribb, Robert. *Historical Atlas of Indonesia.* Honolulu: University of Hawaii Press, 2000.

Cribb, R.B., and Audrey Kahin. *Historical Dictionary of Indonesia.* Lanham, MD: Scarecrow Press, 2004.

Daws, Gavan, and Marty Fujita. *Archipelago: Islands of Indonesia.* Berkeley: University of California Press, 1999.

Draine, Cathie, and Barbara Hall, eds. *Culture Shock: Indonesia.* Portland, OR: Graphic Arts Center Publishing Company, 1991.

Einfield, Jann (ed.). *Indonesia (History of Nations).* San Diego, CA: Greenhaven Press, 2004.

Elson, R.E. *Suharto: A Political Biography.* Cambridge: Cambridge University Press, 2001.

Friend, Theodore. *Indonesian Destinies.* Cambridge, MA: Belknap Press (Harvard University Press), 2003.

Gouri, Mirpuri, and Robert Cooper. *Indonesia (Cultures of the World).* Salt Lake City, UT: Benchmark Books, 2001.

Hooker, M. Barry. *Indonesian Islam: Social Change Through Contemporary Fatawa (Southeast Asia Publications Series).* Honolulu: University of Hawaii Press, 2003.

Jacobs, Judy. Indonesia: *A Nation of Islands (Discovering Our Heritage).* Dillon Press, 1990.

Kingsbury, Damien. *The Politics of Indonesia.* Oxford Press, 2002.

Lansing, J. Stephen. *The Balinese.* Ft. Worth, TX: Harcourt Brace College Publishers, 1995.

Lyle, Gary. *Indonesia.* Philadelphia, PA: Chelsea House Publishers, 1998.

Further Reading

Martin, Fred. *Step into Indonesia.* Portsmouth, NH: Heinemann Educational Books—Library Division, 1998.

Nguyen, Thang G., and Frank-Jurgen Richter. *Indonesia Matters: Diversity, Unity, and Stability in Fragile Times.* Rosebank, South Africa: Times Media, 2003.

Porter, Donald J. *Managing Politics and Islam in Indonesia.* Curzon Press, 2003.

Schwarz, Adam. *A Nation in Waiting: Indonesia's Search for Stability.* Boulder, CO: Westview Press, 1999.

Simkin, T., and R. Fiske. *Krakatau, 1883—The Volcanic Eruption and Its Effects.* Washington, D.C.: Smithsonian Institution Press, 1983.

Simons, G.L. *Indonesia: The Long Oppression.* New York: Palgrave Macmillan, 2000.

Richter, Anne. *Arts and Crafts of Indonesia.* San Francisco, CA: Chronicle Books, 1993.

Ricklefs, M.C. *A History of Modern Indonesia Since C. 1200.* Palo Alto, CA: Stanford University Press, 2002.

Rutherford, Danilyn. *Raiding the Land of the Foreigners: The Limits of the Nation on an Indonesian Frontier.* Princeton, NJ: Princeton University Press, 2002.

Schwarz, Adam, and Jonathon Paris (eds.). *The Politics of Post-Suharto Indonesia.* La Vergne, TN: Lightning Source, 1999.

Taylor, Jean Gelman. *Indonesia: Peoples and Histories.* New Haven, CT: Yale University Press, 2003.

Aceh, 13, 33, 44, 71, 90, 93
Acehnese, 90
Aerowisata International, 80
agriculture, 11, 12, 15, 22, 26, 32, 49, 75, 75–76, 77, 88
Agung, Sultan, 32
Air Paradise International, 80
air pollution, 21–22, 75–76, 96
air travel, 80, 85
Al Qaeda, 10, 44, 53, 88, 96
Albuquerque, Admiral, 29
Allah, 50
Ambon, 28, 55
Ambonese, 56
Amnesty International, 71
angklung, 59
animal domestication, 26
animal life, 14, 19–21, 22, 90
Arabic language, 28–29, 50
Arabs, 27–29, 50, 90
archipelago, 11, 14.
See also islands
architecture, 85
area, 11, 14
art, 29, 57, 87
ASEAN Declaration, 12
Asia, 20, 26, 27
Association of Southeast Asian Nations (ASEAN), 12, 79, 82
Australia, 20, 78, 80
authoritarian government, 40–44, 93

Bahasa Indonesia, 10, 34, 38
Bali, 8–10, 13, 15, 20, 28, 33, 51, 55, 78–79, 80, 83, 87–88, 96
Balinese, 10, 56
Banda Islands, 31
banded krait, 21
Banjarmasin, 33
banking system, 74
Banten, 31, 85
Basa Sunda, 86–87
Batak, 56, 90
Batavia, 32.
See also Jakarta
batik, 57
beaches, 8, 15, 78, 87
Bin Laden, Osama, 53
birds, 20, 21
Bogor, 19
Borneo, 20
Borobudur, 55
Bouraq Indonesia Airlines, 80
Buddhism, 10, 26, 28, 49, 51, 55
Budi Utomo, 34
buffaloes, 20
Buginese, 56
building materials, 77
Bukit Barisan range, 88
bupati (regent), 72
Burma, 25

calderas, 18
Calvinist Dutch Reformed Church, 54
capital city. *See* Jakarta
Celebes, 20
cell phones, 80
Center for Civic Education (Calabasas, California), 94
Central Java, 85, 86
Cerebon, 32
checks and balances, 64, 67, 73
China, 26, 32, 41, 42, 50, 77, 78, 90
Chinese, 56, 59
Christianity, 28, 29, 30, 50, 53–55.
See also Lutheranism; Protestantism; Roman Catholicism
cities, 13, 19, 21–22, 26–27, 49, 85, 87, 96.
See also Jakarta
citizens, role of, 70–71, 73, 94
climate/weather, 8, 11, 14, 18–19, 46, 78–79
coal, 88
Coen, Jan Pieterszoon, 31–32, 85
Cold War, 42
colonial period, 10, 25, 29–35, 50, 63, 74, 90, 92–93
comfort women, 38
communciations, 82
communications, 10, 25, 46, 56, 79–81

Index

Communist Association of the Indies, 34
communists, 39, 40, 41, 42, 54
Confucianism, 55
Constitution of 1945, 38, 53, 55, 63, 64, 66, 67, 68, 70–71, 73
Constitutional Court, 69–70
coral snake, 21
corruption, 12, 23, 32, 40, 43–44, 74, 83, 94–95
coup, 42
culture, 10, 23, 25, 28–29, 30, 35, 46, 48, 48–57, 59–61, 86, 87, 90
Culture/Cultivation System, 32–33
currency, 43, 81–82

dances, 59, 87
Darul Muslim separatist rebellion, 59
Darwin, Charles, 24
Dayaks, 56
deer, 20
deflation, 82
deforestation, 22–23, 95–96
Demak, 28
democracy, 40, 44, 73, 93, 97
  and Pancasila, 38, 48, 63, 93, 94
Denpasar, 87
Department of Religious Affairs, 54
Dewan Perwakilan Daerah (DPD) (Council of Represntatives of the Regions), 67–68
Dewan Perwakilan Rakyat (DPR) (People's Representative Council), 12, 65, 66–67, 68
Diponegoro, Pangeran, 33
Dubois, Eugene, 24–25
Dutch East India Company [Vereenigde Oost-Indische Compagnie (VOC)], 31–32, 85
Dutch East Indies, 31, 33, 36, 39
Dutch language, 10, 38

earthquakes, 11, 15, 92
East Java, 39, 85, 86
East Timor, 11, 30, 40, 43, 44, 93
ebony, 22–23
economy, 12, 23, 43, 44, 74–83, 88
education, 94
Eid al-Adha, 52
Eid al-fitr/Lebanon, 52
elections, 63, 65, 66, 67
elephants, 90
energy resources, 12, 37, 43, 88
English language, 10
environmental problems, 21–23, 75–76, 95–96
equator, 14
ethical period, 33
ethnic cleansing, 55
ethnic conflicts, 55, 56, 59, 74, 77, 83
ethnic groups, 25, 26, 34, 56–57, 59–61, 90, 92
Europeans, 28–29
  See also Great Britain; Netherlands; Portugal
evolution, 24–25
Executive Branch, 12, 64–66
exports, 77–78

fish, 20, 75
Fiske, Richard, 17
five pillars of Islam, 51
Flores, 21, 28
food, 59–60
foreign affairs, 12, 41, 42, 45, 79, 82
foreign debt, 43
foreign investment, 74, 77, 83, 94–95
forests, 21–23, 95–96
France, 30
future, 13, 92–97

garbage, 21–22
Garuda Indonesia Airways, 80
God, and Pancasila, 38, 48, 63, 93
government, 12, 23, 39–45, 49, 50, 53, 54, 62–73, 93, 93–95, 97
Great Britain, 25, 30, 42, 50, 90, 92
green revolution, 75
Gross Domestic Product (GDP), 77
Guided Democracy, 40, 42

Hajj, 52–53
Halmahera, 20, 30
Hasanuddin, Sultan, 32
Hatta, Muhammad, 38, 39, 40
Haz, Hamzah, 96
Henry the Navigator, 29
Hinduism, 10, 26, 27, 28, 49, 51, 55–56, 87, 88
history, 10, 24–35, 25, 36–38, 50, 63, 74, 87–88, 90, 92–93
Hobbs, Thomas, 62
holidays, 48, 52–53
House of People's Representatives, 40
House of Representatives, 40
human rights and liberty, 70–71
humanitarianism, and Pancasila, 38, 48, 63, 93

ice, 11
Ice Age, 20, 25–26
imports, 78
independence, 35, 36, 37, 38, 38–39, 63, 88, 90
India, 26–27, 28, 55, 87, 90
Indian kingdoms, 26–27, 28
Indian Ocean, 11, 28, 84
Indies Party, 34, 35
"Indonesia," and origin of name, 34
Indonesia Democracy Party-Struggle, 44
Indonesian Alliance of Students, 34
Indonesian Nationalist Party (PNI), 35
industrial wastes, 21–22
industry, 12, 75
inflation, 40, 43
insects, 21
instruments, 59
international alliances, 12, 79, 82
International Corruption Watch (ICW), 94
International Monetary Fund (IMF), 74, 82
Internet, 80–81
iodine, 77
Irian Jaya, 11, 44, 71
Islam, 12, 27–29, 35, 39, 49, 50–53, 54–55, 57, 59, 85, 88, 90, 96
Islamic kingdoms, 27–29
islands, 8, 11, 13, 14, 15, 19, 84–88, 90–91, 93
isolation, 25, 46–47, 56, 62
Isra, 52
Istiqlal Mosque, 85

Jakarta, 11, 13, 32, 39, 68, 83, 85, 96
Jakarta Marriott Hotel, 83, 96
Jamaah Islamiyah terrorists, 88
Japan
  exports to, 77
  imports from, 78
  Indonesia occupied by, 25, 36–38, 63, 74, 77, 88, 90, 92
Japanese language, 38
japong, 59
Java, 15, 18, 19, 20, 24–25, 26, 28, 39, 50, 51, 53, 55, 57, 59, 77, 84–87, 96
Java and Dependencies, 30
Java War, 33
Javanese, 56, 57, 59, 60
Javanese language, 10, 34, 57, 86
Jemaah Islamiyah, 96
Jesuits, 54
Judicial Branch, 12, 64, 68–70
Judicial Commission, 68

Ka'ba (Kaaba), 52
kabupaten (regency), 72
Kalemanta, 50
Kalimantan, 33, 57
kaolin, 77
Kelut, 18
Kendal, 21
kendang, 59
Kentucky Fried Chicken, 51
ketuk tilu, 59
king cobra, 21
Komodo, 21
Komodo dragon, 21
Koran (Qu'ran), 12, 50, 53
kota (municipalities), 72
Krakatoa (Krakatau), 17–18

Index

labor force, 49, 75, 77
lahars, 18
lakes, 18
land features, 15, 17–18
languages, 10, 28–29, 30, 34, 38, 57, 86–87
Laskar Jihad, 54–55
Legislative Branch, 12, 64, 65, 66–68
leopards, 90
Liberal Policy, 35
Liberal System, 33
life expectancy, 49
literacy rates, 49
literature, 29
livestock, 75
lizards, 21
local governments, 71–73, 94
location, 11, 14, 18, 19
Locke, John, 62
logging, 22–23
Lombok, 20, 33
Lutheranism, 54

Madura, 33
Madurese, 56
Mahabharata, 57, 59
mahogany, 22–23
Majelis Permusyawaratan Rakyat (MPR) (People's Consultative Asembly), 65, 68
Malacca, 28, 29
Malacca, Strait of, 29, 88
Malay, 59
Malay language, 10, 34. *See also* Bahasa Indonesia
Malay Peninsula, 25, 26, 29, 88
Malays, 56, 90
Malaysia, 11
Malayu, 27
Maluku, 54
Manchuria, 36
manufacturing, 12, 75, 76–79
marriage, 49
marsupials, 20
martial law, 40
Marx, Karl, 35
Mataram, 33
Maulid Nabi, 52–53
Mecca, 52
media, 71, 73, 80, 85
Merpati Nasantara Airlines, 80
metals, 37, 77, 88
military, 40, 42, 42–43, 64, 73
Minangkabau, 56, 90
ministers, 65–66
Miraj, 52
missionaries, 54
Moluccas, 31, 49
monotremes, 20
monsoons, 19
mountains, 11, 14, 15, 19, 46, 48, 78, 88
Muhammad, 28, 50, 52–53
Muhammadiyah (Folowers of Muhammad), 34
Muhammadiyah party, 53
music, 30, 59

Nahdlatul Ulama (NU), 53
Napoleon Bonaparte, 30
national departments, 65–66
national unity, and Pancasila, 38, 48, 56, 63, 73, 93
natural environment, 11, 12, 13, 14–15, 17–23
natural gas, 12, 77
natural hazards, 11, 12, 13, 14, 15, 17–18, 84, 88, 92
natural resources, 12, 37, 43, 77, 83, 88, 93
neighbors, 11, 46
Netherlands (Holland), 10, 25, 30–35, 36, 37, 38, 39, 40, 50, 54, 63, 74, 88, 90, 92
New Guinea, 15
New Order, 42–43
nighttime, 18–19
noise pollution, 96

oil, 12, 37, 43, 77, 88
Organization of Petroleum Exporting Countries (OPEC), 43, 77

Pacific Ocean, 11, 84
Pacific Rim, 12
Palembang, 26–27
Paleolithic age, 25
Pancasila, 38, 43, 44, 48, 53, 56, 63, 73, 93–94, 97
Papua, 71
Papua New Guinea, 11
Pelita Air Service, 80
people, 10, 56–57, 59–61. *See also* ethnic groups
per capita income, 49
Perlak, 28, 50
Philippines, 25
phosphate, 77
pigs, 20
piracy, 29
Pithecanthropus erectus (Java man), 24–25
plains, 15
plant life, 14, 19–20, 22, 90
plantations, 32, 75
political instability, 12–13, 39–40, 44, 59, 73, 74, 75, 76–77, 83, 93
political parties, 34–35, 40, 42, 44, 50, 53, 70
political prisoners, 71
Polo, Marco, 28, 50, 90
popular culture, 85
population, 10–11, 22, 45, 48–49, 82, 93
population density, 11, 48–49, 96
population growth, 49, 96
ports, 26–27, 28, 29, 31, 85
Portugal, 25, 29–30, 31, 50, 54, 74, 90, 92
Portuguese language, 30
postal service, 80
precipitation, 11, 14, 15, 18, 19
prehistory, 24–26
president, 12, 38–44, 54, 63, 64–66, 67, 68, 69–70
President for Life, Sukarno as, 40–42
Project Citizen, 94
Protestantism, 10, 49, 54
provinces, 72–73, 85–86, 90
Puncak Jaya, 15
python, 21
radios, 80
Raffles, Thomas Stamford, 30
railways, 80
rainfall, 19
Ramadan, 51–52
Ramayana, 57
Regional People's House of Representatives, 72–73
religion, 10, 25, 26, 27, 28, 29, 30, 48, 49–50, 51, 53–56, 87, 88
religious conflict, 54–55, 77, 83
Republic of Indonesia, 12, 39, 63–64
rhinoceroses, 20, 90
rice, 26, 75, 82
Ring of Fire, 11, 12, 15
roads, 15, 80
Roman Catholicism, 10, 49, 50, 54
rubber, 37
Rupiah, 43, 81–82

Sambal Goreng Ayam, 60
sandalwood, 22–23
Sanskrit language, 26
Sarekat Islam (Islamic Union), 34, 35
Sari Club, 10, 44, 83, 88
Sasak, 10
Sasaks, 56
Saudi Arabia, 52, 53
science, 29
sea snakes, 21
separatist movements, 12–13, 39–40, 44, 59, 73, 74, 77, 83, 90, 93, 94
service industries, 75, 77, 79
settlement patterns, 33
Seventeen Gentlemen, 31
sewage, 21–22
Sharia law, 12, 90
shipbuilding, 30
Simkin, Thomas, 17
Singapore, 30, 77, 78
slash-and-burn techniques, 22, 75–76
slavery, 30, 33
snakes, 21
snow, 11, 14, 15, 19
social justice, and Pancasila, 38, 48, 63, 93–94

Index

Soetomo, 34
South Korea, 77, 78
Soviet Union, 41
Spice Islands, 31, 49, 59
spice trade, 26, 29, 31–32, 50, 85
spices, 59–60, 75
Srivijaya Kingdom, 26–27, 28, 55
Sufi Islam, 50–51, 53
Suharto, 42–44, 71, 93, 94, 96
Sukarno, 34–35, 37, 38–42, 44, 48, 53, 54, 56, 63, 93
Sukarnoputri, Megawati, 44
Sulawesi, 21, 33, 54, 57
sulfur, 77
Sumatra, 11, 26, 28, 39, 50, 57, 77, 88, 90
Sumbawa Island, 15, 17
Sunda, 59
Sunda Kelapa Harbor, 85
Sunda Kingdom, 87
Sundanese, 10, 56, 57, 59
Supreme Court, 12, 68–69

Taiwan, 77
Tambora Volcano, 13, 17
Taoism, 55
tapirs, 90
teak, 22–23
technology, 46, 48, 79–81, 82
tectonic plates, 15
telephones, 80
televisions, 80
temperatures, 18–19
Ternate
  revolt on, 29–30
  Sulton of, 29
terrorism, 9–10, 44, 53, 73, 74, 83, 88, 96
Thailand, 25, 55
tigers, 20, 90
tobacco, 30
tourism, 8–10, 15, 78–79, 80, 83, 87, 88
transportation, 10, 15, 25, 46, 56, 79–80, 82
trees, 19, 21–23, 95–96
tropical rain forests, 19, 22, 90, 96
tropics, 11, 18
Trunojoyo, Prince, 32
tsunamis, 15, 17–18

Ubud, 87
unemployment, 75
United Nations, 41, 42
United States, 39, 40–41, 42, 77, 78, 96
United States Agency for International Development (USAID), 94
United States Central Intelligence Agency (U.S. CIA), 42

vice-president, 39, 40, 44, 64, 68, 96
vipers, 21
volcanoes, 11, 13, 15, 17–18, 84, 88, 92

Wahhabism, 53
Wahid, Abdurrahman 'Gus Dur,' 44
walikota (mayor), 72
Wallace, Alfred R., 20
Wallace's Line, 20
water features, 18, 21
water pollution, 21–22, 96
wayang goleks, 59
wayang (shadow theater), 57, 59
West Java, 31, 57, 85, 86–87
West New Guinea, 40
West Papua, 28, 33
Western Java, 39
winds, 19
women, 49
World Bank, 94
World War II, 36, 38, 39, 63, 88

Yogyakarta, 86

Picture Credits

page:

9: © Lucidity Information Design, LLC
13: New Millennium Images
16: © Lucidity Information Design, LLC
22: New Millennium Images
27: New Millennium Images
37: © Bettmann/CORBIS
41: © Bettmann/CORBIS
47: New Millennium Images
52: AFP/New Millennium Images
69: Reuters Photo Archive/NMI
72: EPA/WEDA/AP/Wide World Photos
76: New Millennium Images
78: AP/Wide World Photos
81: EPA/WEDA/AP/Wide World Photos
86: KRT/New Millennium Images
89: New Millennium Images
95: New Millennium Images

Cover: New Millennium Images

About the Contributors

DOUGLAS A. PHILLIPS is a lifetime educator, writer, and consultant who has worked and traveled in more than 80 countries on six continents. During his career he has worked as a middle-school teacher, curriculum developer, author, and as a trainer of educators in many countries around the world. He has served as the President of the National Council for Geographic Education, and he has received the Outstanding Service Award from the National Council for the Social Studies, together with numerous other awards. He, his wife Marlene, and their three children, Chris, Angela, and Daniel, have lived in South Dakota and Alaska, but his daughter is now in Texas while he, his wife, and his two sons reside in Arizona, where he writes and serves as an educational consultant for the Center for Civic Education. He has traveled in Indonesia and understands the importance of this incredible Southeast Asian nation.

CHARLES F. ("FRITZ") GRITZNER is Distinguished Professor of Geography at South Dakota University in Brookings. He is now in his fifth decade of college teaching and research. During his career, he has taught more than 60 different courses, spanning the fields of physical, cultural, and regional geography. In addition to his teaching, he enjoys writing, working with teachers, and sharing his love for geography with students. As consulting editor for the MODERN WORLD NATIONS series, he has a wonderful opportunity to combine each of these "hobbies." Fritz has served as both President and Executive Director of the National Council for Geographic Education and has received the Council's highest honor, the George J. Miller Award for Distinguished Service. In March 2004, he won the Distinguished Teaching award from the American Association of Geographers at their annual meeting held in Philadelphia.